Alexander Schwab

Cooperative Control of Networked Vehicles

Logos Verlag Berlin

λογος

Bibliografische Information der Deutschen Nationalbibliothek

Die Deutsche Nationalbibliothek verzeichnet diese Publikation in der
Deutschen Nationalbibliografie; detaillierte bibliografische Daten sind
im Internet über http://dnb.d-nb.de abrufbar.

ISBN 978-3-8325-5549-8

Logos Verlag Berlin GmbH
Georg-Knorr-Str. 4, Geb. 10,
D-12681 Berlin, Germany

Tel.:   +49 (0)30 / 42 85 10 90
Fax: +49 (0)30 / 42 85 10 92

http://www.logos-verlag.de

Cooperative Control of Networked Vehicles

Dissertation zur Erlangung des Grades eines

DOKTOR-INGENIEURS

der Fakultät für Elektrotechnik und Informationstechnik an der

RUHR-UNIVERSITÄT BOCHUM

ALEXANDER SCHWAB

geboren in Alma-Ata

Bochum, 2022

1. Gutachter: Prof. Dr.-Ing. Jan Lunze
             Ruhr-Universität Bochum

2. Gutachter: Prof. Dr.-Ing. Daniel Görges
             Technische Universität Kaiserslautern

Tag der mündlichen Prüfung: 1. Juli 2022

Acknowledgement

This thesis is the result of several years of research at the Institute of Automation and Computer Control at Ruhr-University Bochum, Germany. These years would not have been such a great experience without the following people to whom I am deeply grateful.

Firstly, I would like to thank my supervisor Professor Jan Lunze who has significantly shaped my understanding of control theory. In our discussions, he has given me a direction and, simultaneously, the freedom to explore problems off the beaten path. He taught me to take a step back and think about a problem without getting lost in details.

Special thanks goes to my colleagues Michael Schwung, Philipp Welz, Markus Zgorzelski, Kai Schenk, Christian Wölfel, Marc Wissing, Fabian Schneider, Sven Bodenburg, Andrej Mosebach, Daniel Vey, Melanie Schuh and René Schuh who have made this demanding time enjoyable. I will always remember our discussions about our research and, more importantly, the number of buttons of a radio controlled clock as well as how to get to the moon. I am glad that these people have become my friends.

Furthermore, I am grateful to Kerstin Funke, Andrea Marschall and Susanne Malow who have taken care of all organisational matters and always stood by with advice and assistance. I would also like to thank my student assistants Florian Littek and Nora Lindner who helped me to put my ideas into practice.

Last but not least, I would like to express my heartfelt gratitude to my mother Anna, my brother Arthur and his wife Kristina together with their lovely kids Niklas, Emilia and Lennart as well as all my other friends. You are the port to which i can return again and again, no matter what the future has in store for me.

Bochum, July 2022   Alexander Schwab

Abstract

The present thesis concerns the cooperative control of networked vehicles. Autonomous driving is a topic that is currently being discussed with great interest from researches, vehicle manufactures and the corresponding media. Future autonomous vehicles should bring the passengers to their desired destination while improving both safety and efficiency compared to current human-driven vehicles. The inherent problem of all vehicle coordination tasks is to guarantee collision avoidance in every situation. To this end, autonomous vehicles have to share information with each other in order to perform traffic manoeuvres that require the cooperation of multiple vehicles.

By providing analysis tools and design methods, the theory of networked control systems adds to the development of vehicle controllers with which a collaboration of multiple networked vehicles shows the desired cooperative behaviour. The general idea, which is pursued in this thesis, is to impose requirements (e. g. collision avoidance) on the overall behaviour of the controlled vehicles. Based on these requirements, properties of the controlled vehicles on the one hand and properties of the communication network on the other hand are derived. The local controller of each vehicle and the structure of the communication network have to be designed to meet specific design objectives so that the combination of the controlled vehicles and the communication network achieves the desired overall behaviour.

This procedure is applied to the control of networked vehicles in different ways. At first, scenarios that can be encountered in real world traffic are considered. As a fundamental problem, vehicle platooning is studied extensively which describes the task of arranging a set of vehicles so that they drive with a common velocity and a prescribed distance. Using the procedure described above, local design objectives are derived that have to be satisfied by the vehicle controllers. In particular, it is shown that the vehicles have to be externally positive to achieve collision avoidance. However, the problem of rendering a feedback loop externally positive has been unsolved so far. The present thesis solves the design problem for a general class of linear vehicle models based on a state feedback. The proposed controller structure and parametrisation is shown to achieve the desired behaviour with information that can be gathered via local sensors or the communication network.

As an abstraction from real traffic scenarios, swarms of mobile systems are considered. The main difference between swarming and traffic problems is the freedom of movement since a swarm of mobile systems is not bound to move on prescribed roads. An important implication of this circumstance is that a communication structure that has been appropriate in the beginning might become unsuited for the control task due to the relative movement of the mobile systems. To solve this problem, this thesis proposes to use the Delaunay triangulation as a switching communication structure of a networked system. To this end, it will be shown that the Delaunay triangulation can be maintained by the mobile systems with distributed algorithms.

The task to be solved by the swarm of mobile agents is to achieve a distance-based formation. Due to the free motion of the mobile systems on a driving surface, nonlinear local controllers are applied that perform a gradient descent along an artificial potential field that is composed of relative potential functions. It will be shown how to choose and parametrise the relative potential functions so that the combination of the Delaunay triangulation network structure and the local controllers solve the control task while guaranteeing collision avoidance.

All proposed methods are evaluated with a laboratory plant which consists of several mobile robots that are tracked by a camera system. The experiments verify the theoretical results under consideration of the following aspects. The robots have properties and limitations (e. g. control input bounds, dead band for small inputs, wheel slippage, etc.) that are not considered in the models that describe the nonholonomic kinematics. Nevertheless, the proposed control techniques are shown to be real-time capable and to achieve collision avoidance in all considered scenarios.

Contents

Introduction to cooperative control of networked vehicles

1.1 Networked control systems

The networking of controlled subsystems will play an increasingly important role in technical automation in the future. Instead of a central control system, each subsystem will have a local controller that is connected to other controllers via a communication network. The exchange of information ensures that a common task can be solved cooperatively. The distributed structure also allows for easy scalability as subsystems are added to or removed from the network.

The present thesis concerns the control of many vehicles that should cooperate with each other to achieve common or individual goals while guaranteeing collision avoidance. From a control-theoretic point of view, the collaboration of multiple subsystems, which represent the controlled vehicles, is represented as a *multi-agent system*. The structure of such a system is presented in Fig. 1.1.

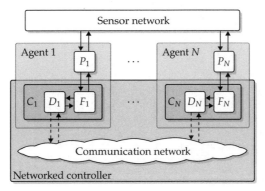

Fig. 1.1: Networked control structure of a multi-agent system

The controlled subsystems are referred to as the *agents* $i = 1, 2, \ldots, N$ and are composed of multiple units. The plant P_i represents the physical behaviour of a vehicle. In order to achieve a desired behaviour of the agents, the local controller C_i determines the input of the plant. To this end, the controller is itself composed of the feedback unit F_i that executes an appropriate control algorithm and the communication and decision unit D_i that is connected to the communication network and collects all information that is needed by F_i to operate.

Due to the connection and interaction of the local controllers, a multi-agent system is said to have a *networked control structure*. The communication and decision units D_i send and receive information via the communication network. The information is not limited to physical quantities as state variables but might also contain intentions of an agent and instructions to other agents. Thus, for the sake of safety and efficiency, the agents are able to make decisions based on their available information and cooperate with other agents if that is required to achieve the individual or common control aim.

In addition to the digital communication network, the agents are coupled via a sensor network which represents the measurement of physical quantities that concern other agents. In the context of networked vehicles, the sensor network represents the coupling of the vehicles via the inter-vehicle distance as an example. The inclusion of physical interactions in the model structure allows for introducing the notion of *cognition* which describes the ability of the agents to sense their environment or other agents that are not connected to the communication network. In contrast to classical physical interconnections as in power plants or processing systems that are coupled statically in close proximity, the sensor network of mobile agents is flexible just like the communication network. That is, the coupling structure can change dynamically.

1.2 Goal of this thesis

The present thesis considers multiple vehicles that should cooperate with each other in order to solve different coordination problems. The fields of application in this thesis to be presented in the next section range from real traffic scenarios to abstract formation problems. In the considered traffic scenarios, the vehicles have individual aims (e. g. to reach a destination) or common goals (e. g. to form a platoon or a formation). The fundamental goal in each situation is to guarantee *collision avoidance*.

> The goal of this thesis is to develop methods for the cooperative control of networked vehicles that achieve collision avoidance and to test them in different traffic scenarios.

This task is considered under the circumstance that there is no global coordinator. Consequently, the vehicles have to cooperate with each other by sharing information via the communication network.

Way of solution

The idea of the design of networked control systems in this thesis is illustrated in Fig. 1.2. First, it is considered which behaviour the overall system should have to achieve an overarching control aim. The result should be a list of requirements on the networked control system. Based on these requirements, it is then elaborated which properties the agents and the communication network should have. The challenge is to find local properties that can be implemented by the local feedback F_i on the one hand and a network structure that can be determined and maintained locally by the communication and decision unit D_i on the other hand so that the combination of the agents and the communication network satisfies the requirements on the overall system.

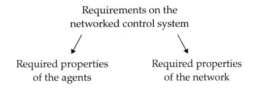

Fig. 1.2: Illustration of the controller design process

1.3 Fields of application

1.3.1 Autonomous vehicles

A common example concerning the control of networked systems are connected and automated vehicles (CAV). Autonomous driving is a highly discussed topic and is expected to improve traffic efficiency and safety [139]. Modern vehicles come with several driver assistance systems that are able to maintain a safety distance to the predecessor or to perform autonomous lane changes.

Vehicle platooning

Vehicle platooning describes the problem of arranging a set of vehicles in a line so that the inter-vehicle distances match a predefined (possibly velocity dependent) spacing. Adaptive cruise control (ACC) is a driver assistance system that should solve this task using sensor data only. Figure 1.3 shows the basic structure of an ACC platoon with vehicles that are equipped with radar sensors to measure the distance and a controller to adapt the velocity of each vehicle with the aim that a desired distance is achieved and all vehicles travel with a common velocity asymptotically. To benefit from reduced fuel consumption due to exploitation of slipstream, the distance has to be small which leads to the fact that a driver in an automated vehicle is not able to react on a disturbance in time to prevent an imminent collision. Thus, the cruise controller has to guarantee collision avoidance for an arbitrary length of the platoon.

Fig. 1.3: Platoon with adaptive cruise control

An extension to ACC is cooperative adaptive cruise control (CACC) which uses the communication network in addition to the distance sensors. This approach allows for considering additional information from vehicles further downstream. Thus, the controlled vehicles can react faster on changes of the reference velocity due to traffic conditions or disturbances.

Vehicle merging

Vehicle merging can be interpreted as an extension of platooning which concerns the coordination of multiple platoons so that two or more platoons merge into a single one. Such techniques are required to pass highway on-ramps or lane reductions due to construction sites automatically. Commercially available approaches to the control of autonomous vehicles are currently based on the intensive use of sensor technology and estimation methods to predict the behaviour of other vehicles. Thus, lane changing assistants have to wait for a gap to occur coincidentally since there is no cooperation between vehicles. By introduction of a communication network, this limitation can be resolved.

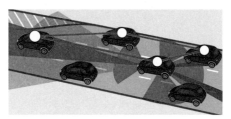

Fig. 1.4: Connected and automated vehicles passing through a lane reduction [86]

Figure 1.4 shows an example with multiple automated vehicles which are coupled in two ways. In addition to the measurement of the distance to the predecessor as required for platooning, the vehicles are equipped with sensors to detect the environment and other vehicles as illustrated by the cones. Furthermore, the local vehicle controllers can communicate with each other via wireless communication links (illustrated by the yellow edges) to exchange information and to perform cooperative manoeuvres if necessary.

With adaptive cruise controllers, the communication system is used only when a cooperation of multiple vehicles is mandatory to perform a specific manoeuvre, for example, when there is no gap on the target lane that is sufficiently large. The vehicles on the target lane then receive a request from the merging vehicle and generate a gap cooperatively for the merging vehicle to steer in.

1.3.2   Swarms of mobile systems

Motivated by nature, the collective behaviour of flocks of birds or ants serves as a model for swarms of mobile systems. The goal is to create patterns or formations that serve an overarching goal without the use of a centralised coordinator.

Fig. 1.5: A swarm of networked robots

Figure 1.5 shows a swarm of robots that is connected via communication links. The networked control of mobile agents has a wide area of applications to solve coverage, logistics or satellite control problems. The fundamental common task in these applications is to create desired formations by networked controllers.

In contrast to control problems concerning connected and automated vehicles, swarms of mobile agents are not bound to move on fixed roads but rather are able to move freely on a plane. This degree of freedom in the movement has to be considered by the networked controller to achieve the formations while guaranteeing collision avoidance.

1.4 Open questions

For the solution of the presented applications, two essential questions have to be examined.

- **Design of the local controllers:** What properties do the controlled vehicles have to have so that they can be combined in any way and in any number to achieve an overarching control aim?

- **Design of the communication structure:** When does a vehicle have to communicate with which other vehicles so that all vehicles can avoid collisions together?

These questions have been answered partially for vehicle platoons in [84, 88]. It has been shown that the vehicles in a platoon have to be *externally positive* in order to guarantee collision avoidance. However, the design of externally positive control loops is not solved in general [6]. The present thesis starts at that point and will address and solve the design of adaptive cruise controllers that render the controlled vehicle externally positive.

When considering vehicle platoons, the communication structure is often given as a path graph. In this thesis, the general question of an appropriate communication structure will hence be addressed for abstract swarms of vehicles that are not bound to prescribed paths as in Fig. 1.5. The vehicles are free to move on the surface and relatively to one another. Thus, the a communications structure that has been purposeful in the beginning can become unsuitable as the vehicles move along their trajectories. In order to find an appropriate communication structure that adapts to the current geometrical configuration of the vehicles, a proximity network based on the Delaunay triangulation will be studied as the communication structure of a networked system with mobile agents.

The following selection of references is based on the aspects that have been addressed in this section.

1.5 Literature overview

1.5.1 Networked control systems

In the field of networked systems, the most fundamental problem is the *consensus problem* with simple integrator dynamics where all agents should agree on a common steady state. It is shown that the graph-theoretic interpretation of the communication structure must include a spanning tree to achieve this goal [107]. The stability of networked systems is often studied in combination with communication effects as packet losses or delays [147].

An extension to higher-dimensional systems with the goal of bringing all control variables onto a common trajectory defines the *synchronization problem* which is often studied for subsystems with identical dynamics [125]. For the synchronization of heterogeneous systems, the internal-reference principle was presented in [81] and [82] which shows that the subsystems to be synchronised must necessarily have common dynamics.

1.5.2 Control of networked vehicles

Vehicle platooning has been shown to potentially increase efficiency and safety in daily traffic [139]. There are several recent papers that study energy-optimal adaptive cruise control from an ecological point of view [59, 60, 74].

One of the earliest works on vehicle platooning considered centralised optimal control techniques [73]. However, due to the decentralised character of traffic problems, decentralised methods to satisfy safety requirements are necessary for practical applications. To this end, methods of the control of networked systems have been applied to the control of connected and automated vehicles [40, 121, 122].

An important and fundamental requirement considering the control of vehicles is collision avoidance. Thus, in [132], the concept of *string stability* has been introduced and it has been shown that the string stability condition cannot be satisfied using distributed controllers in combination with a constant distance policy [127]. Thus, the idea of a time-dependent inter-vehicle distance was analysed, for example, in [68] which was applied in various theoretical and practical studies [64, 99, 103, 114, 146]. All these publications use a specific vehicle model with integrating dynamics and apply weak string stability to ensure safety requirements by shaping the closed-loop transfer function so that its magnitude is bounded. However, as has been proven in [84], the stricter property of external positivity has to be imposed on the controlled vehicles to guarantee collision avoidance. External positivity implies strong string stability because an externally positive vehicle is also $\mathcal{L}_\infty$ string stable [37].

The experimental results presented in [46, 98] have shown that commercially available vehicles are not externally positive at that time. Thus, the design of adaptive cruise controllers that render the controlled vehicles externally positive requires further investigations.

1.5.3 Externally positive systems

There are several books and articles on positive linear systems, e. g. [39, 61, 79, 118] which are based on the theory of positive matrices that goes back to Perron and Frobenius [24].

Despite being a less restrictive property, the literature on *externally positive systems* is far less comprehensive since external positivity is hard to characterise by the parameters of a system. To this end, some effort has been conducted based on the observation that externally positive systems have a nonovershooting step response. An interesting observation was published in [67] giving an upper bound for the number of extrema of the step response in terms of the location of the zeros of a system. This work motivated several authors to present necessary or sufficient conditions that avoid overshooting step responses [57, 78, 117, 133]. However, some of the conditions in these publications are either too complicated to be used to design control loops or are too restrictive with respect to the location of the eigenvalues [57]. Simpler sufficient conditions for external positivity were published in [58, 117] which are based purely on the configuration of the poles and zeros of a system.

There are several fields of application of positive systems in economics, population analysis, probabilistic models and compartmental systems [49]. In the monograph [39], a whole chapter is dedicated to the application of externally positive systems. Some recent publications showed that (external) positivity of the subsystems is a helpful property in the control of networked systems which simplifies the analysis of the stability and the consensus under constraints such as uncertain or probabilistic communication structures [25, 27, 86, 138]. Furthermore, external positivity is a property of controlled vehicles that can guarantee collision avoidance under well defined conditions [37, 84].

Even though there is a wide area of applications, there are only a few publications on the design of externally positive systems since the design problem is rather difficult. This is due to the fact that external positivity is characterised by a nonnegative impulse response which can be hardly determined analytically. Thus, the general design problem is unsolved [6]. In [134], a compensator is presented that renders a system externally positive but the contribution does not provide a constructive way for the feedback design problem. For multiple-input multiple-output systems, there are two publications that try to shape the eigenvectors of the controlled system so that each output is represented by one mode of the system which renders it externally positive. In [42], a linear matrix inequality was deduced that can be solved with optimal control techniques and a similar approach in [126] uses the modal synthesis to shape the eigenvectors. However, these approaches suffer from the limitation that the design method is only applicable to systems with unit relative degree which restricts the field of application. Some constructive sufficient conditions for external positivity based on a factorisation of the transfer function were published in [56, 58, 117].

1.5.4 Design of communication structures

The design of communication structures for networked control systems is a problem that has not been solved yet in general [86]. However, some effort for special control applications has been conducted, for example, in [88] where the design of a communication structure for platoons with CACC is elaborated. The results are based on a delay measure introduced in [83] that allows for transforming the design problem of a network structure into an algebraic problem. Furthermore, there are approaches that find an optimal communication topology with respect to the overall system performance by solving mixed-integer semidefinite programs [44, 48]. However, these methods are centralised in the design of the network topology and the network is fixed once it is determined.

The present thesis proposes the Delaunay triangulation as the network structure which should be maintained decentrally by the agents along their trajectories. The Delaunay triangulation is a well studied structure with a rich literature in the fields of computational geometry, geodesy, computer graphics, data clustering or spatial discretisation for finite element methods. It is the dual of the Voronoi diagram [52, 116] and is named after its inventor who addressed this topic in the original paper [35] where he defined his triangulation by its specific property that the unique circle through any three points connected as a triangle does not contain any other point in its interior. The Delaunay triangulation can be constructed for a given Voronoi diagram or directly for a given set of points [72]. A popular method for generating the Delaunay property is the Lawson flip which allows for adjusting two adjacent triangles if the empty-circumcircle property is violated [70, 71]. Conditions to test this property are given in [45, 52, 116].

The Delaunay triangulation belongs to the class of *proximity graphs* [95]. The structure of such graphs is determined by the relative position of its vertices. There are publications on networked systems that seek consensus using proximity graphs, e. g. [21] and [31] where the authors used r-disk graphs and integrators to model the agents. Some other applications use the Delaunay triangulation to evaluate and improve the placement of the vertices, e. g. ad-hoc networks or coverage control problems which seek an optimal distribution of the vertices in a given area [32, 75, 96]. The Delaunay triangulation has also been applied for formation control of multiple unmanned aircrafts [28] or path planning for autonomous vehicles [20]. In the field of computational geometry, the maintenance of the Voronoi diagram or the Delaunay triangulation of moving points was discussed in [19, 41, 130, 148] with centralised methods. The application of the Delaunay triangulation as a network structure for multi-agent systems requires the maintenance to be distributed so that it can be performed cooperatively by the agents which will be investigated in this thesis.

1.5.5 Swarms of mobile agents

A famous publication by Vicsek et al. [141] presented a study of particles that change their direction of movement by applying nearest neighbour rules. They showed that all particles agree on a direction under specific parameter values. This observation shows that it is possible to coordinate groups of subsystems without a superior coordinator which is desirable in the distributed control of mobile agents.

Distance-based formations as an example have been studied in various publications with linear (double) integrator models or nonholonomic models [36, 144]. In the related field of *flocking*, the agents should form a group in which the agents also align their velocity vectors so that the group travels together in a consensual direction [54, 106].

In order to solve such formation control problems, gradient-based control schemes with artificial potential fields (sometimes referred to as edge-tension functions, Lyapunov-like functions or barrier functions) are predestined due to the target-oriented influence on the agents [51, 97]. Since the gradient-method is well suited for the control of vehicle swarms, it will be applied in this thesis.

Collision avoidance as a fundamental requirement is often achieved with unbounded control inputs which result from the application of diverging potential functions [38, 93, 94, 119, 135–137]. While, theoretically, these schemes guarantee provable collision avoidance, they are not reliable in practice since control inputs are always bounded.

The application of artificial potential fields for formation control and collision avoidance on linear integrator models was discussed in several publications, e. g. in [38, 93, 106, 108, 143]. It is possible to extend these results to nonholonomic kinematics by application of a feedback linearisation [51, 119]. The direct application of the potential field method on nonholonomic kinematics was, for example, discussed in [50, 77, 94, 112, 113].

Control application with swarms of mobile agents usually apply a switching communication structure due to the individual movement of the agents. Related literature has shown that for the standard consensus protocol with integrator dynamics, consensus is achieved if the switching communication structure is always strongly connected and balanced which is true for any connected undirected graph [109]. The requirement has been relaxed so that the switching communication does not always have to possess a spanning tree but rather frequently enough [120]. These results also have been discussed and extended in the context of flocking in [54, 135–137] with nearest-neighbour rules.

By introduction of a switching communication graph, the resulting control system is a hybrid system with continuous dynamical states of the agents and discrete states of the communication graph. There are different tools to evaluate the stability of hybrid systems, e. g. multiple Lyapunov functions [29, 30, 34, 76] or a single common Lyapunov function that is applied to all discrete states [91] or nonsmooth analysis [129, 137]. However, these methods are difficult to apply as there are no systematic ways to construct the corresponding Lyapunov functions. Furthermore, it has been shown that there are switched linear systems that are stable but there does not exist a common Lyapunov function since the existence of a common quadratic Lyapunov function is only sufficient for stability. Some effort to prove stability of flocking algorithms with switching networks or nearest neighbour interactions with linear integrator dynamics can be found in [135–137].

1.6 Contributions of this thesis

The present thesis will address both the design of local controllers that implement desired properties of the agents and the choice of an appropriate communication network. The results (lemmas, theorems and algorithms) of the author are presented in boxes. Results that are taken from the literature are presented without a box marked with the corresponding reference. The main contributions of this thesis are described in the following.

1.6.1 Design of externally positive feedback loops

It has been shown that external positivity is a valuable property for various applications. However, the design of controllers that render a given plant externally positive is an unsolved problem [6]. This observation is due to the fact that external positivity of a linear system is characterised by its impulse response which is hardly determinable in dependence of the controller parameters analytically [100].

This thesis will summarise known conditions for external positivity in order to discuss approaches to solve the problem of designing controllers that result in an externally positive feedback loop. Theorem 3.4 reveals that plants with multiple eigenvalues equal to zero are not positively stabilisable.

Thus, in order to design adaptive cruise controllers, a state-feedback approach is proposed to solve the design problem for a general class of vehicle models. Sufficient conditions on the closed-loop eigenvalues that achieve externally positive closed-loop dynamics are given by Theorem 4.4. A design procedure for adaptive cruise controllers, which guarantee collision avoidance by rendering the controlled vehicles externally positive, will be developed and summarised by Algorithm 4.2. The effectiveness of the proposed algorithm is verified in an experimental environment in Section 7.3.

1.6.2 Maintenance of a proximity communication network

Swarms of mobile agents as an abstraction of real traffic problems require the communication structure to adjust to the current geometrical configuration of the agents. The Delaunay triangulation is proposed as a good choice of a communication structure for mobile agents due to the following reason. It has always a spanning tree and it has the important property that the pairwise closest agents are always connected which is an important property concerning the collision avoidance [95].

This thesis proposes to use the Delaunay triangulation as the communication network for the control of mobile agents. To this end, it will been shown that it is possible to maintain the Delaunay triangulation by the decentralised Algorithm 5.4 that is executed by all agents. The proposed algorithm is based on an efficient local characterisation of the Delaunay triangulation given by Theorem 5.1.

1.6.3 Collision avoidance in swarms of mobile systems

As the third theoretical contribution, this thesis studies the combination of a switching Delaunay triangulation network structure with local controllers that should achieve a distance-based formation. In contrast to the related literature that uses fixed communication structures and unbounded control inputs, this thesis will present a way to guarantee collision avoidance under consideration of control input limitations by the application of the Morse potential function [101] which has been first applied to control problems in [128].

With the parametrisation of the Morse potential function by Theorem 6.1, collision avoidance will be shown by Theorem 6.3 if the Delaunay triangulation communication structure is maintained during the transition of the agents. The result is summarised in Algorithm 6.1 which gives explicit instruction to apply the proposed controller in combination with the Delaunay triangulation network.

1.6.4 Experimental evaluation of the theoretical results

The theoretical results of this thesis are evaluated with laboratory experiments using differentially driven robots. The experimental results complement the theoretical results with further insights into practical aspects of the control applications.

In order to test the proposed adaptive cruise controller, a path tracking controller is developed that steers the robot to a prescribed path as shown by Theorem 7.1. The combination of simultaneous lateral and longitudinal control is studied with vehicle platoons that should follow a circular path or merge into a single platoon in case of a lane reduction. The experiments verify that collision avoidance in the platoon is achieved with external positive dynamics in contrast to platoons with $\mathcal{L}_2$ string stable vehicles that end up in a collision.

The remaining experiments consider swarms of robots that should form a distance-based formation and, as an extension, the transition problem is studied that describes the task of reaching an individual destination point while avoiding collisions with other robots. The experiments reveal that the proposed algorithms that maintain the Delaunay triangulation are real-time capable. Furthermore, the characterisation of the Morse potential achieves collision avoidance in the laboratory environment which verifies the theoretical results.

1.7 Structure of this thesis

The main results of this thesis are presented in Chapters 3 – 7. In the following, a brief overview of the content of each chapter is given.

Chapter 2 gives an overview of fundamental control-theoretic methods that will be applied in this thesis.

Chapter 3 introduces externally positive systems and summarises known properties and characteristics of such systems. Furthermore, it will be discussed for which classes of systems there exists a controller that renders the feedback loop externally positive or not.

Chapter 4 addresses the platooning problem. Several requirements are given that represent the desired behaviour of the overall platoon. Based on these requirements, design objectives are developed that should be achieved by local controllers. The design problem is then solved for a general class of vehicle models. In particular, a method to design externally positive vehicles using a state feedback is proposed.

Chapter 5 concerns the maintenance of a proximity communication network based on the Delaunay triangulation. First, the Delaunay triangulation is introduced and local conditions are developed to characterise the current network from the perspective of an agent. Based on these conditions, a method to test and adjust the current network is developed so that it represents a Delaunay triangulation after each iteration.

Chapter 6 solves the task of forming distance-based formations using gradient-based control techniques. It will be shown how to parametrise Morse potential functions that guarantee collision avoidance while respecting control input limitations. The results are extended to the case that a switching Delaunay triangulation network is applied and it will be shown under which conditions the collision avoidance is preserved.

Chapter 7 presents measurements that verify the theoretical results of the previous chapters using mobile robots. To this end, a lateral path controller is developed that complements the longitudinal platooning controller. Furthermore, the proposed swarming methods are evaluated with robots that are free to move on the driving surface.

Chapter 8 summarises the presented results.

Preliminaries and fundamentals

2

This chapter summarises fundamental models, methods and concepts of control theory that are relevant for the subsequent chapters.

2.1 Notation

Matrices and vectors are typeset in boldface (e. g. $\boldsymbol{A}$ or $\boldsymbol{x}$). The norm of a vector is denoted by $\|\boldsymbol{x}\|$ which represents the Euclidean norm if not specified otherwise. The determinant of a matrix is denoted by $|\boldsymbol{A}|$. For a scalar argument, $|a|$ denotes the absolute value. Furthermore, inequalities with vectors as $\boldsymbol{x} \geq 0$ are understood element wise.

Functions of time t are denoted by lower-case letters (e. g. $g(t)$) and the corresponding function in the Laplace domain by the same upper-case letters (e. g. $G(s)$). The Laplace transform is symbolized by $\circ\!\!-\!\!\bullet$ as in the example $g(t) \circ\!\!-\!\!\bullet G(s)$. The limit of a function is abbreviated as

$$\lim_{t \to \infty} g(t) = g(\infty).$$

The functions $\sigma(t)$ and $\delta(t)$ denote the Heaviside (unit step) function and the Dirac (impulse) distribution, respectively.

An overset exclamation mark as in the examples

$$h(t) \overset{!}{\geq} 0, \quad t \geq 0, \qquad\qquad h(0) \overset{!}{=} 0$$

indicates a condition that *should* be satisfied by an appropriate choice of parameters.

The gradient of a multivariable scalar function $f(\boldsymbol{x}) = f(x_1, x_2, \ldots, x_n)$ is denoted by $\nabla f(\boldsymbol{x})$ and is defined as the column vector of the partial derivatives

$$\nabla f(\boldsymbol{x}) := \left(\frac{\partial f(\boldsymbol{x})}{\partial x_1} \quad \frac{\partial f(\boldsymbol{x})}{\partial x_2} \quad \cdots \quad \frac{\partial f(\boldsymbol{x})}{\partial x_n} \right)^{\mathrm{T}}.$$

The gradient with respect to a part of the argument, e. g. $\tilde{\boldsymbol{x}}^{\mathrm{T}} = \begin{pmatrix} x_1 & x_2 \end{pmatrix}$ is denoted by

$$\nabla_{\tilde{\boldsymbol{x}}} f(\boldsymbol{x}) := \left(\frac{\partial f(\boldsymbol{x})}{\partial x_1} \quad \frac{\partial f(\boldsymbol{x})}{\partial x_2} \right)^{\mathrm{T}}.$$

Sets are denoted by calligraphic letters as $\mathcal{E}$. The empty set is denoted by $\emptyset$ and the cardinality of a set is given by $|\mathcal{E}|$ which represents the number of elements of a set.

2.2 Description of linear time-invariant systems

2.2.1 Description in the time domain

This thesis considers linear time-invariant systems denoted by Σ. The input and the output of Σ are denoted by the scalar signals $u(t)$ and $y(t)$, respectively. The dynamical behaviour of Σ is described by the n-th order ordinary differential equation

$$\Sigma: \frac{d^n y(t)}{dt^n} + a_{n-1} \frac{d^{n-1} y(t)}{dt^{n-1}} + \ldots + a_1 \frac{dy(t)}{dt} + a_0 \, y(t)$$
$$= b_q \frac{d^q u(t)}{dt^q} + b_{q-1} \frac{d^{q-1} u(t)}{dt^{q-1}} + b_1 \frac{du(t)}{dt} + \ldots + b_0 \, u(t) \tag{2.1}$$

with constant coefficients $a_i, (i = 0, 1, \ldots, n-1)$ and $b_i, (i = 0, 1, \ldots, q)$. A system Σ that satisfies $n \geq q$ is called *proper* or *strictly proper* if $n > q$ holds. A system that is not proper has no technical realisation. In a diagram, a linear system is represented by a block as illustrated in Fig. 2.1.

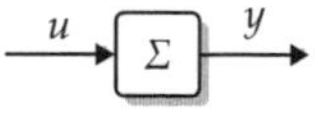

Fig. 2.1: Graphic representation of a linear system

For $n > q$, the differential equation (2.1) can be cast in the state-space representation

$$\Sigma: \begin{cases} \dot{x}(t) = A\,x(t) + b\,u(t), & x(0) = x_0 \\ y(t) = c^{\mathsf{T}} x(t) \end{cases} \tag{2.2}$$

with $x(t)$ denoting the *state* of Σ [89]. The state-space model (2.2) is abbreviated as $\Sigma = (A, b, c^{\mathsf{T}})$. The parameters A, b and c^{T} can be constructed for a given set of coefficients $a_i, (i = 0, 1, \ldots, n-1)$ and $b_i, (i = 0, 1, \ldots, q)$ and characterise the dynamical input-output behaviour completely. Note that the parameters A, b and c^{T} are not unique, i.e. there is an infinite number of sets of parameters that describe Σ equivalently.

The input-output behaviour of Σ is alternatively characterised by the impulse response

$$g(t) = c^{\mathsf{T}} e^{At} b,$$

which is the output of Σ if the input is the Dirac impulse: $u(t) = \delta(t)$.

2.2.2 Description in the frequency domain

The state-space model (2.2) is a time-domain representation of Σ. By introduction of the Laplace transforms $Y(s) \bullet\!\!-\!\!\circ y(t)$ and $U(s) \bullet\!\!-\!\!\circ u(t)$, an equivalent frequency domain representation of the input-output behaviour is given by

$$\Sigma : \ Y(s) = G(s)U(s)$$

with the *transfer function* $G(s)$ that is determined by the elements of (2.2) as

$$G(s) = c^{\mathsf{T}}(sI - A)^{-1}b.$$

The transfer function $G(s)$ is the Laplace transform of the impulse response $g(t)$ of Σ. For a given set of parameters $a_i, (i = 0, 1, \ldots, n-1)$ and $b_i, (i = 0, 1, \ldots, q)$ (cf. (2.1)), the transfer function can be constructed as

$$G(s) = \frac{b_q s^q + b_{q-1}s^{q-1} + \cdots + b_1 s + b_0}{s^n + a_{n-1}s^{n-1} + \cdots + a_1 s + a_0}.$$

The frequencies s that render the numerator or the denominator equal to zero are called *transmission zeros* s_{0i} or *poles* s_i of Σ, respectively. If they are known, the transfer function is equivalently given as the factorisation

$$G(s) = \kappa \frac{\prod_{i=1}^{q}(s - s_{0i})}{\prod_{i=1}^{n}(s - s_i)}$$

with some scalar constant κ. All poles s_i of $G(s)$ are furthermore also eigenvalues λ_i of A.

2.3 Fundamentals of control theory

2.3.1 Controllability

Consider a strictly proper plant $\Sigma = (A, b, c^{\mathsf{T}})$. The question of whether Σ can be influenced so that a specified control aim is achieved (e. g. bring the output $y(t)$ to a desired value) is characterised by the following property.

Definition 2.1 (Controllability [90]). A system Σ is said to be *controllable* if there is an input $u(t), (t \in [0, t_e])$ that moves the state $x(t)$ from any initial state x_0 to any given final state $x(t_e)$ in final time t_e.

In order to test a given model for the controllability property, there are different necessary and sufficient conditions [90]. The criterion proposed by Kalman in [62] uses the *controllability matrix* which is defined as

$$S := \begin{pmatrix} b & Ab & \cdots & A^{n-1}b \end{pmatrix}. \tag{2.3}$$

Theorem 2.1 (Controllability [62, 90]). *A system $\Sigma = (A, b, c^T)$ is controllable if and only if the controllability matrix (2.3) is regular:*

$$\text{rank } S = n \tag{2.4}$$

Since the controllability of a system is determined by A and b, the pair (A, b) is said to be controllable if (2.4) holds true. Note that the property *observability* is closely related to controllability and is determined by the pair (A, c^T). It characterises whether the initial state x_0 of a system can be determined given the input $u(t)$ and output $y(t)$ in a closed time interval $[0, t_e]$. The conditions for observability are dual to the conditions for controllability and, hence, not repeated here. The interested reader is referred to [90].

Controllability, poles and zeros

The controllability of a system is linked to the poles and zeros as shown in this section. For a strictly proper system $\Sigma = (A, b, c^T)$, the Rosenbrock system matrix is defined as

$$P(s) = \begin{pmatrix} sI - A & -b \\ c^T & 0 \end{pmatrix},$$

which allows for computing the zeros of a system as follows. All frequencies s_{0i} that satisfy

$$\det P(s_{0i}) = 0$$

are called *invariant zeros* of Σ. An invariant zero s_{0i} that is also an eigenvalue λ_i of the system matrix A is a *decoupling zero*. Decoupling zeros do not appear as poles in the transfer function $G(s)$ and, hence, are neither controllable nor observable. Conversely, a system that has no decoupling zeros is controllable and observable. In this case, $\Sigma = (A, b, c^T)$ is said to be a *minimal realisation* of $G(s)$. Furthermore, an invariant zero s_{0i} that satisfies $G(s_{0i}) = 0$ is a transmission zero as already discussed in Section 2.2.2.

Control canonical form

A strictly proper system Σ that is controllable can be cast in the state-space representation

$$\Sigma : \begin{cases} \dot{x}_R(t) = \begin{pmatrix} 0 & 1 & 0 & \cdots & 0 \\ 0 & 0 & 1 & \cdots & 0 \\ \vdots & \vdots & \vdots & \ddots & \vdots \\ 0 & 0 & 0 & \cdots & 1 \\ -a_0 & -a_1 & -a_2 & \cdots & -a_{n-1} \end{pmatrix} x_R(t) + \begin{pmatrix} 0 \\ 0 \\ \vdots \\ 0 \\ 1 \end{pmatrix} u(t), \quad x_R(0) = x_{R0} \\ y(t) = \begin{pmatrix} b_0 & b_1 & \cdots & b_q & 0 & \cdots & 0 \end{pmatrix} x_R(t) \end{cases} \tag{2.5}$$

with the coefficients a_i and b_i of the differential equation (2.1). The control canonical form (2.5) is relevant for pole assignment techniques using state feedbacks as shown in the following section.

2.3.2 State feedback loops and pole assignment

In the standard control loop, the output $y(t)$ is fed back to the controller in order to determine the control input $u(t)$. A state feedback instead uses the whole state $\boldsymbol{x}(t)$ to determine the input according to

$$u(t) = -\boldsymbol{k}^\mathsf{T}\,\boldsymbol{x}(t) + v\,w(t) \tag{2.6}$$

with $w(t)$ denoting the reference signal. The constants $\boldsymbol{k}$ and v denote the feedback gain and the prefilter, respectively. The combination of (2.2) with (2.6) results in the model

$$\bar{\Sigma} : \begin{cases} \dot{\boldsymbol{x}}(t) = \left(\boldsymbol{A} - \boldsymbol{b}\boldsymbol{k}^\mathsf{T}\right)\boldsymbol{x}(t) + \boldsymbol{b}v\,w(t), \quad \boldsymbol{x}(0) = \boldsymbol{x}_0 \\ y(t) = \boldsymbol{c}^\mathsf{T}\boldsymbol{x}(t) \end{cases} \tag{2.7}$$

of the state feedback loop that is shown in Fig. 2.2.

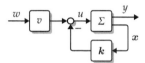

Fig. 2.2: State feedback structure

The prefilter is chosen according to

$$v = -\frac{1}{\boldsymbol{c}^\mathsf{T}\left(\boldsymbol{A} - \boldsymbol{b}\boldsymbol{k}^\mathsf{T}\right)^{-1}\boldsymbol{b}}$$

so that $\bar{\Sigma}$ has a unit static reinforcement. The feedback gain $\boldsymbol{k}$ has to be chosen so that the feedback loop is asymptotically stable in addition to other objectives imposed on the desired closed-loop dynamics. There are different ways to determine $\boldsymbol{k}$. In the following, an approach that is applied in this thesis is presented which allows for choosing the closed-loop eigenvalues.

Ackermann's formula

The closed-loop system matrix

$$\bar{\boldsymbol{A}} = \boldsymbol{A} - \boldsymbol{b}\boldsymbol{k}^\mathsf{T} \tag{2.8}$$

of (2.7) is influenced significantly by the choice of the feedback gain $\boldsymbol{k}$. The researcher J. Ackermann proposed a way to determine the feedback gain $\boldsymbol{k}$ that places the closed-loop eigenvalues as desired [18].

The idea is as follows. Suppose that the state-space model parameters of the plant are given in the control canonical form (2.5). The closed-loop system matrix (2.8) is then given by

$$
\bar{A}_R = \begin{pmatrix}
0 & 1 & 0 & \cdots & 0 \\
0 & 0 & 1 & \cdots & 0 \\
\vdots & \vdots & \vdots & \ddots & \vdots \\
0 & 0 & 0 & \cdots & 1 \\
-\bar{a}_0 & -\bar{a}_1 & -\bar{a}_2 & \cdots & -\bar{a}_{n-1}
\end{pmatrix}
$$

with $\bar{a}_{i-1} = a_{i-1} + k_{Ri}$. Thus, the feedback gain, which implements a given set of coefficients $\bar{a}_{i-1}, (i = 1, 2, \ldots, n)$ of the closed-loop characteristic polynomial, is determined by

$$
k_{Ri} = \bar{a}_{i-1} - a_{i-1}, \quad i = 1, 2, \ldots, n. \tag{2.9}
$$

Note that a feedback gain determined with (2.9) will correspond to the feedback of the state of the control canonical form:

$$
u(t) = -k_R^T x_R(t) + v\, w(t).
$$

In order to generalise the result (2.9) for any plant model that is not in control canonical form, the feedback gain has to be transformed according to

$$
k^T = k_R^T T_R^{-1}
$$

with the transformation matrix

$$
T_R = \begin{pmatrix}
s_R^T \\
s_R^T A \\
s_R^T A^2 \\
\vdots \\
s_R^T A^{n-1}
\end{pmatrix}, \tag{2.10}
$$

where s_R^T denotes the last row of the inverse controllability matrix (2.3):

$$
s_R^T = \begin{pmatrix} 0 & 0 & \cdots & 0 & 1 \end{pmatrix} S^{-1}. \tag{2.11}
$$

The combination of (2.9) with (2.10) and application of the Cayley–Hamilton theorem finally results in Ackermann's formula

$$
k^T = \begin{pmatrix} \bar{a}_0 & \bar{a}_1 & \bar{a}_2 & \cdots & \bar{a}_{n-1} & 1 \end{pmatrix}
\begin{pmatrix}
s_R^T \\
s_R^T A \\
s_R^T A^2 \\
\vdots \\
s_R^T A^{n-1} \\
s_R^T A^n
\end{pmatrix}. \tag{2.12}
$$

For a given set of desired closed-loop eigenvalues $\bar{\lambda}_i, (i = 1, 2, \ldots, n)$, the closed-loop coefficients $\bar{a}_i, (i = 0, 1, \ldots, n-1)$ are obtained with the characteristic polynomial

$$\bar{P}(\lambda) = \sum_{i=0}^{n} \bar{a}_i \, \lambda^i = \prod_{i=1}^{n} (\lambda - \bar{\lambda}_i) \tag{2.13}$$

of the closed-loop system with $\bar{a}_n = 1$ by convention. The design procedure is as follows.

1. Choose the desired closed-loop eigenvalues $\bar{\lambda}_i, (i = 1, 2, \ldots, n)$.

2. Determine the coefficients $\bar{a}_i, (i = 0, 1, \ldots, n-1)$ of the polynomial (2.13).

3. Determine the feedback gain k with Ackermann's formula (2.12).

The presented procedure requires the pair (A, b) to be controllable which can be seen in (2.11). In order to calculate s_R^T, the inverse of the controllability matrix has to be determined which possible if and only if the plant is controllable due to Theorem 2.1.

2.4 Networked control systems

2.4.1 Description of networked control systems

This section gives a very brief introduction to networked control systems. The interested reader is referred to the textbook [86] for a thorough elaboration on this topic.

Consider a set of N agents with identical dynamics described by

$$P_i : \begin{cases} \dot{x}_i(t) = A \, x_i(t) + b \, u_i(t), & x_i(0) = x_{i0} \\ y_i(t) = c^T x_i(t), \end{cases} \qquad i = 1, 2, \ldots, N.$$

The goal is to bring the outputs $y_i(t)$ of all agents to some synchronous trajectory $y_s(t)$:

$$\lim_{t \to \infty} |y_s(t) - y_i(t)| = 0, \quad i = 1, 2, \ldots, N. \tag{2.14}$$

A networked control system that achieves (2.14) for all initial states $x_{i0}, i = (1, 2, \ldots, N)$ is said to be *synchronised*.

In order to achieve asymptotic synchronisation, the agents are connected with each other via a communication network. The network structure is represented by a graph $\mathcal{G} = (\mathcal{V}, \mathcal{E})$ with $\mathcal{V} = \{1, 2, \ldots, N\}$ and $\mathcal{E}$ denoting the set of vertices and the set of edges, respectively. If there is a directed edge $(j \to i) \in \mathcal{E}$, then agent j is said to be a *neighbour* of agent i, i.e. agent i receives information from agent j. The set of all neighbours of an agent is defined as

$$\mathcal{N}_i := \{j \mid (j \to i) \in \mathcal{E}\}.$$

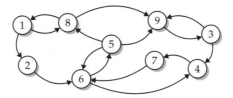

Fig. 2.3: Graph-theoretic representation of a networked system

An example graph is shown in Fig. 2.3.

By introduction of the adjacency matrix $A_\mathcal{G} = (a_{ij})$ with the elements

$$a_{ij} = \begin{cases} 1 & \text{if } (j \rightarrow i) \in \mathcal{E} \\ 0 & \text{otherwise,} \end{cases}$$

a networked controller that is able to achieve (2.14) is given by

$$F_i : \ u_i(t) = - \sum_{j \in \mathcal{N}_i} a_{ij} \left(y_i(t) - y_j(t) \right), \quad i = 1, 2, \ldots, N.$$

An algebraic representation of the network structure can be obtained as follows. Consider the degree matrix $D = \text{diag } d_i$ with the elements

$$d_i = \sum_{j=1}^{N} a_{ij} = |\mathcal{N}_i|.$$

The Laplacian matrix is then defined as

$$L := D - A_\mathcal{G}$$

and has the following properties.

- It is singular since all row sums are zero.

- All eigenvalues have a nonnegative real part.

- The graph that is represented by a Laplacian matrix L has a spanning tree if all eigenvalues except for one are nonzero.

The last property of the Laplacian matrix is very interesting since it is possible to test whether there is a root vertex from which there is a path to all other vertices by calculating the eigenvalues of the Laplacian matrix. As example, consider the matrix

$$
L = \begin{pmatrix}
1 & 0 & 0 & 0 & 0 & 0 & 0 & -1 & 0 \\
-1 & 1 & 0 & 0 & 0 & 0 & 0 & 0 & 0 \\
0 & 0 & 1 & 0 & 0 & 0 & 0 & 0 & -1 \\
0 & 0 & -1 & 2 & 0 & -1 & 0 & 0 & 0 \\
0 & 0 & 0 & 0 & 1 & -1 & 0 & 0 & 0 \\
0 & -1 & 0 & 0 & -1 & 3 & -1 & 0 & 0 \\
0 & 0 & 0 & -1 & 0 & 0 & 1 & 0 & 0 \\
-1 & 0 & 0 & 0 & -1 & 0 & 0 & 2 & 0 \\
0 & 0 & -1 & 0 & -1 & 0 & 0 & -1 & 3
\end{pmatrix}
$$

for the graph in Fig. 2.3.

A spanning tree is a necessary condition for the asymptotic synchronisation (2.14) of a set of agents since, otherwise, the information does not spread through the whole network and, hence, it is not possible for the agents to agree on a common synchronous trajectory. Further necessary and sufficient conditions for the synchronisation of a multi-agent system are given in [86].

2.4.2 Delay measure

When considering networked system, it is interesting to characterise the time lag with which some information distributes among the network. The *delay measure* is an abstract model of a dynamical system that describes the response of an agent to some new input (or information). The delay is defined as follows.

Definition 2.2 (Delay [86]). The *delay* of a linear system P_i is defined as

$$
\Delta_I := \int_0^\infty \left(k_{si} - h_i(t) \right) dt, \tag{2.15}
$$

where $k_{si} = h_i(\infty)$ and $h_i(t)$ denote the static reinforcement and the step response of P_i, respectively.

The definition (2.15) of the delay is illustrated in Fig. 2.4. It represents the area between the final value and the step response. Thus, a small delay corresponds to a response that reaches the final value quickly, and, conversely, a system with a large delay needs more time to settle on the final value.

For a given transfer function $G_i(s)$, the delay can be determined explicitly without solving the integral (2.15) as follows.

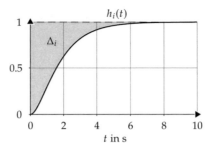

Fig. 2.4: Delay of system with $k_{si} = 1$

Lemma 2.1 (Determination of the delay [86]). *For an asymptotically stable system P_i with the transfer function $G_i(s)$, the delay (2.15) is determined by*

$$\Delta_i = -\lim_{s \to 0} \frac{dG_i(s)}{ds}.$$

The delay measure can be used to characterise or design communication structures that enable the agents to synchronise in a desired time span [83, 85]. Furthermore, it has been shown that the delay measure is an important property of vehicle platoons with time-headway spacing [84].

Externally positive systems

3

This chapter summarises known properties of externally positive systems and conditions on external positivity that can be used to design externally positive control systems by an appropriate choice of a feedback. As the main contribution of the author, it is shown in Section 3.3 that for plants with multiple integrators, it is not possible to render the closed-loop system externally positive using the standard control configuration. For externally positive plants on the other hand, it is straightforward to find an appropriate controller.

3.1 Introduction to externally positive systems

3.1.1 The notion of positive systems

There are systems with state variables that are naturally nonnegative due to the underlying physics (the filling level of a tank, the temperature measured on the Kelvin scale or the concentration of dissolved substances) or nonnegative by construction (probabilistic processes represented by Markov models or comparison systems used for stability analysis [80]). The notion of positive systems was introduced to characterise these kind of systems with nonnegative state variables that are encountered typically in biology, chemistry or in the process industry [27].

A linear system

$$\Sigma : \begin{cases} \dot{x}(t) = A\,x(t) + b\,u(t), & x(0) = x_0 \\ y(t) = c^{\mathsf{T}}x(t) \end{cases} \tag{3.1}$$

is called *positive* if its state $x(t)$ and its output $y(t)$ are nonnegative when starting from an initial condition $x_0 \geq 0$ subject to any nonnegative input $u(t)$ [39]. The conditions on positive systems are rather strong, because it is required that all elements of the state-space representation $\Sigma = (A, b, c^{\mathsf{T}})$ are nonnegative with the exception of the diagonal entries of A. For several applications it is sufficient that the output of a system is nonnegative even if some state variables are negative and, hence, this chapter addresses a relaxed version of positive systems which is the class of *externally positive* systems defined as follows.

Definition 3.1 (Externally positive system [39]). A linear system (3.1) is called *externally positive* if its zero-state response is nonnegative for any nonnegative input:

$$x_0 = 0, \quad u(t) \geq 0, \quad t \geq 0 \quad \Longrightarrow \quad y(t) \geq 0, \quad t \geq 0.$$

An externally positive system can be characterised by its impulse response $g(t)$ as shown by the following theorem.

Theorem 3.1 (Externally positive system [39, 61]). *A linear system (3.1) is externally positive if and only if its impulse response is nonnegative:*

$$g(t) = c^T e^{At} b \geq 0, \quad t \geq 0. \tag{3.2}$$

As a direct consequence of Theorem 3.1, externally positive systems have a monotonically increasing step response

$$h(t) = \int_0^t g(\tau) \, d\tau,$$
$$\dot{h}(t) = g(t) \geq 0, \quad t \geq 0.$$

Thus, the step response of an externally positive system does not overshoot its final value. This property is very useful for various control applications.

3.1.2 Externally positive feedback loops

There are plants that have to be made externally positive by a feedback to satisfy performance or safety specifications. Two typical examples are a milling machine, which will ruin a workpiece if its transitions show an overshoot, or a serial interconnection of systems (e. g. a vehicle platoon) which must be externally positive to achieve string stability and collision avoidance [37, 84].

Consider the closed-loop system

$$\bar{\Sigma} : \; Y(s) = \bar{G}(s) W(s)$$

with the reference $w(t) \circ\!\!-\!\!\bullet W(s)$ and the closed-loop transfer function $\bar{G}(s)$. The goal of the design procedure is to satisfy the objectives

(i) Asymptotic stability
(ii) Set-point following
(iii) External positivity

by choosing an appropriate controller structure and controller parameters. The special aspect of the design objective is to render the impulse response of the feedback loop nonnegative

$$\bar{G}(s) \bullet\!\!-\!\!\circ \bar{g}(t) \geq 0, \quad t \geq 0, \tag{3.3}$$

which is necessary and sufficient for external positivity due to Theorem 3.1. Unfortunately, the objective **(iii)** is difficult to satisfy with a design procedure since there is no easy way to represent the impulse response $\tilde{g}(t)$ of the closed-loop system $\tilde{\Sigma}$ in terms of the controller parameters in order to satisfy the inequality (3.3) [100].

This chapter will address the following two fundamental problems:

- Given a linear plant model Σ, check whether there exists a controller so that (3.3) holds true.

- Given a linear plant model Σ, find a controller so that the objectives **(i)–(iii)** are satisfied.

A review of the relevant literature reveals that these problems are unsolved in general [6]. The purpose of this chapter is to summarise available conditions for external positivity as a fundament for the analysis in Chapter 4 where the design problem is solved for vehicles.

3.2 Further conditions on external positivity

3.2.1 Necessary conditions

In order to characterise external positivity, further conditions are presented in the following. First, the dominant eigenvalue of a system will be defined followed by three fundamental necessary conditions for external positivity.

Definition 3.2 (Dominant eigenvalue [134]). Let $\{\lambda_1, \lambda_2, \ldots, \lambda_n\}$ denote the set of (self conjugate) eigenvalues $\lambda_i = \sigma_i + j\omega_i$ of a system. The eigenvalue with the largest real part

$$\lambda_{\text{dom}} = \max_i \text{Re}\{\lambda_i\}$$

is called the *dominant eigenvalue*. If there are multiple eigenvalues with equal real part, the eigenvalue with the highest algebraic multiplicity is the dominant eigenvalue. Furthermore, if there are multiple eigenvalues with equal real part and multiplicity, the eigenvalue with the smallest imaginary part is the dominant eigenvalue.

It is well known in system theory that a complex conjugate dominant pole pair introduces sinusoidal components in the impulse response causing the sign to switch infinitely often and rendering the impulse response partially negative [134]. The following lemma is a consequence of this observation.

Lemma 3.1 (Real dominant eigenvalue [133]). *An externally positive system has a real dominant eigenvalue:*

$$\text{Im}\{\lambda_{\text{dom}}\} = 0.$$

Another necessary condition on the magnitude plot of an externally positive system can be deduced from the condition (3.2) as shown by the following lemma.

> **Lemma 3.2** (Bounded magnitude plot). *The magnitude plot of an externally positive system is bounded by its static gain:*
>
> $$\sup_{\omega} |G(j\omega)| \le G(0).$$

Proof. The transfer function is the Laplace transform of the impulse response $g(t)$

$$G(s) = \int_0^\infty g(t)\,e^{-st}\,dt.$$

Explicit calculation of the magnitude yields

$$|G(j\omega)| \le \int_0^\infty |g(t)|\,\underbrace{\left|e^{-j\omega t}\right|}_{=\,1}\,dt = \int_0^\infty g(t)\,dt = G(0)$$

using the triangle inequality for integrals, nonnegativity of the impulse response (3.2) and the fact that the integral of $g(t)$ is equal to the static gain which proves the lemma. ∎

The following lemma is a direct consequence of Lemma 3.2 since a real zero raises the magnitude plot.

Lemma 3.3 (Zero location [133]). *All real zeros μ_i, $(i = 1, 2, \ldots, q)$ of an externally positive system are located to the left of the dominant eigenvalue:*

$$\mu_i < \lambda_{\text{dom}}, \quad i = 1, 2, \ldots, q.$$

Lemmas 3.1–3.3 present conditions for external positivity that can be satisfied in a straightforward way in the controller design process. Unfortunately, these conditions are only necessary and, thus, a system that satisfies them is not guaranteed to be externally positive. The impulse response has to be calculated numerically to verify whether it is nonnegative as shown by the following example.

Example 3.1 (Not externally positive system). The transfer function

$$G(s) = \frac{375\,(s + 1.4)(s + 1.6)}{7\,(s + 1)(s + 4)(s + 5)(s + 6)} \tag{3.4}$$

has the dominant real eigenvalue $\lambda_{\text{dom}} = -1$ with two zeros located to the left of the dominant eigenvalue and the magnitude plot is bounded by the static gain $G(0) = 1$ as shown in Fig. 3.1 (a). Thus, the conditions of Lemmas 3.1–3.3 are satisfied. However, the system is not externally positive as revealed by the impulse response in Fig. 3.1 (b) even though all poles and zeros are real.

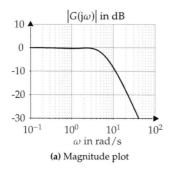
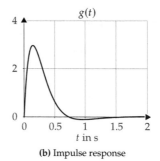

(a) Magnitude plot

(b) Impulse response

Fig. 3.1: Properties of the example system (3.4)

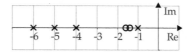

Fig. 3.2: Poles (crosses) and zeros (circles) of system (3.4)

Discussion. This example shows that the zeros of a system are of significant importance concerning the question whether said system is externally positive. Obviously, the two consecutive zeros close to the dominant eigenvalue shown in Fig. 3.2 are responsible for rendering the impulse response negative. This observation has been studied in various literature references (e. g. [57, 78, 133]) resulting in different pole-zero patterns that achieve external positivity. The simplest one is an alternating pole-zero pattern. □

3.2.2 Sufficient conditions

The necessary conditions presented in the last section describe characteristic properties of externally positive systems. However, they are not useful in the design process of externally positive feedback loops. To this end, some known sufficient conditions for external positivity are presented in this section. A fundamental result published in [67] presents an upper bound that establishes a connection between the real zeros of a system and the number of zero crossings of its impulse response.

Theorem 3.2 (Impulse response zero crossings [67]). *Consider a linear system* (3.1) *with real poles and zeros and, furthermore, assume that the condition of Lemma* 3.3 *is satisfied. The number of zero crossings of the impulse response* $g(t)$ *is bounded from above by* $\tilde{q} - m$, *where* $\tilde{q}$ *is the number of zeros located between two adjacent poles and m is the number of occurrences that an odd number of zeros is located between two adjacent poles.*

In Example 3.1, the transfer function (3.4) has $\tilde{q} = q = 2$ zeros located in the same pole bracket (cf. Fig. 3.2) and, hence, $m = 0$. Thus, its impulse response has possibly two zero crossings and, indeed, it has exactly two zero crossings as depicted in Fig. 3.1 (b).

Theorem 3.2 can be used to formulate sufficient conditions on external positivity by ensuring that the upper bound of zero crossings is equal to zero. Particularly noteworthy are the results presented in [58, 117] where sufficient conditions to characterise externally positive systems were presented that consider a factorisation of the transfer function

$$G(s) = \kappa \prod_{i=1}^{\tilde{n}} G_i(s), \quad \tilde{n} \le n, \quad \kappa > 0, \tag{3.5}$$

which represents a series connection of $\tilde{n}$ subsystems.

Definition 3.3 (Positively decomposable system). A linear system represented by the transfer function $G(s)$ is said to be *positively decomposable* if there exists a factorisation (3.5) that satisfies

$$G_i(s) \bullet\!\!-\!\!\circ g_i(t) \ge 0, \quad t \ge 0, \quad i = 1, 2, \dots, \tilde{n}. \tag{3.6}$$

If it is possible to decompose a transfer function $G(s)$ according to (3.6), it can be concluded that the system with the transfer function $G(s)$ is externally positive as shown by the following lemma.

Lemma 3.4 (External positivity by positive decomposition [117]). *A positively decomposable system according to Definition 3.3 is externally positive.*

Proof. The impulse response of (3.5) can be written as the chain of convolutions

$$g(t) = \kappa \, (g_1 * g_2 * \cdots * g_{\tilde{n}})(t).$$

Since all elements of the convolution are nonnegative due to (3.6) and the convolution of nonnegative functions is again nonnegative, $g(t)$ is also nonnegative. ∎

For systems with real poles $s_i, (i = 1, 2, \dots, n)$ and real zeros $s_{0i}, (i = 1, 2, \dots, q)$, a simple factorisation of $G(s)$ with first-order subsystems is given by

$$G_i(s) = \begin{cases} \dfrac{s - s_{0i}}{s - s_i}, & i = 1, 2, \dots, q \\[2ex] \dfrac{1}{s - s_i}, & i = q + 1, q + 2, \dots, n. \end{cases} \tag{3.7}$$

The elements of (3.7) correspond to the impulse responses

$$g_i(t) = \begin{cases} (s_i - s_{0i})\, e^{s_i t}, & i = 1, 2, \dots, q \\[1ex] e^{s_i t}, & i = q + 1, q + 2, \dots, n, \end{cases}$$

which are nonnegative if and only if the condition

$$s_{0i} \leq s_i, \quad i = 1, 2, \ldots, q \tag{3.8}$$

is satisfied. That is, the zero s_{0i} has to lie to the left of the associated pole s_i in the complex plane. These results are summarised in the following theorem.

Theorem 3.3 (Pole-zero configuration [58, 117]). *Consider a transfer function $G(s)$ with real poles and zeros. If there exists a factorisation (3.5) and (3.7) so that the condition (3.8) is satisfied, the system represented by $G(s)$ is externally positive.*

A factorisation with the factors (3.7) assumes real poles and zeros and, hence, is only applicable to systems that satisfy this assumption. However, the approach (3.6) of finding a positive decomposition is also applicable to systems with complex conjugate poles. To this end, factors of higher order, e. g.

$$G_i(s) = \frac{1}{(s + s_i)\big((s + \sigma_i)^2 + \omega_i^2\big)} \tag{3.9}$$

with complex conjugate poles are introduced in [58] where it is stated that the transfer function (3.9) corresponds to a nonnegative impulse response if and only if $\sigma_i \leq s_i \leq 0$ holds true. In [133], it is furthermore shown that also complex conjugate zeros (even in the right half plane) are possible in externally positive systems. However, there is no complete collection of factors to characterise all externally positive systems with the factorisation approach. Nevertheless, Theorem 3.3 is very useful in the design of externally positive feedback loops and will be applied in Chapter 4 to design externally positive vehicles.

3.3 On the design of externally positive feedback loops

3.3.1 Design problem

This section addresses the problem of finding a controller for a given plant so that the resulting feedback loop is externally positive. The standard control loop depicted in Fig. 3.3 will be considered. The plant and the controller are represented by the transfer functions $G(s)$ and $K(s)$, respectively. The transfer function of the closed-loop system is given by

$$\bar{G}(s) = \frac{G(s)\, K(s)}{1 + G(s)\, K(s)}. \tag{3.10}$$

As described in the introduction of this chapter, the questions to be addressed in the following are whether there exits a controller $K(s)$ and, if that is the case, how to find the controller structure and parameters that satisfy

$$\bar{G}(s) \bullet\!\!-\!\!\circ \bar{g}(t) \geq 0, \quad t \geq 0. \tag{3.11}$$

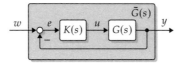

Fig. 3.3: Standard control loop

These questions are largely open and will not be answered in general here. Nevertheless, the results in the following sections answer these questions partially for special cases that can be used as a fundament for further research.

3.3.2 Existence of a solution

The difficulty in designing externally positive control loops lies in the fact that the nonnegativity of the impulse response according to Theorem 3.1 is an inequality that has to be satisfied with conditions on the parameters of the closed-loop system represented by $\bar{g}(t)$ which is a difficult task since there is no way to characterise the impulse response analytically [39, 100]. The following definition characterises the existence of a solution to the design problem.

Definition 3.4 (Positive stabilisation). A plant represented by the transfer function $G(s)$ is said to be *positively stabilisable* by an output feedback (Fig. 3.3) if there exists a controller $K(s)$ that renders the closed-loop system externally positive according to (3.11) and asymptotically stable.

The following consideration shows that there are classes of transfer functions that are *not* positively stabilisable. To this end, consider a plant with a transfer function that is composed according to

$$G(s) = \frac{1}{s^m}\tilde{G}(s) \tag{3.12}$$

with $m > 1$ poles equal to zero. Furthermore, assume that there are no zeros in the open loop located in the origin of the complex plane that cancel the poles equal to zero, i.e. the plant and the controller satisfy

$$\tilde{G}(0) \neq 0, \quad K(0) \neq 0. \tag{3.13}$$

The closed-loop transfer function (3.10) is then given by

$$\bar{G}(s) = \frac{\tilde{G}(s)K(s)}{s^m + \tilde{G}(s)K(s)} \tag{3.14}$$

with $\bar{G}(0) = 1$ due to (3.13). The following theorem shows that it is not possible to render (3.14) externally positive regardless of the choice of the controller $K(s)$.

> **Theorem 3.4** (Integrating plants). *A plant (3.12) with $m > 1$ integrators (poles equal to zero) is not positively stabilisable, i. e. there does not exist any controller $K(s)$ that renders the closed-loop impulse response $\tilde{g}(t)$ nonnegative in the standard control loop (Fig. 3.3) under condition (3.13).*

Proof. Let $\bar{h}(t)$ denote the step response of a closed-loop system with the the transfer function (3.14). The control error

$$e(t) := w(t) - y(t) = \sigma(t) - \bar{h}(t)$$

over the course of the step response $\bar{h}(t)$ corresponds to

$$e(t) \circ\!\!-\!\!\bullet\ E(s) = \frac{1}{s} - \frac{1}{s}\,\tilde{G}(s) = \frac{s^{m-1}}{s^m + \tilde{G}(s)\,K(s)}$$

in the frequency domain. According to the final value theorem, the integral of the control error $e(t)$ is given by

$$\lim_{t \to \infty} \int_0^t e(\tau)\,d\tau = \lim_{s \to 0} E(s)$$

$$= \lim_{s \to 0} \frac{s^{m-1}}{s^m + \tilde{G}(s)\,K(s)}$$

$$= \frac{0}{\tilde{G}(0)K(0)} = 0. \tag{3.15}$$

Equation (3.15) shows that the areas enclosed by $\bar{h}(t)$ and the final value $\bar{h}(\infty) = 1$ cancel each other exactly. Consequently, $\bar{h}(t)$ has decreasing sections that correspond to negative intervals in the impulse response independently of the controller which proves the theorem. ∎

Example 3.2 (Double integrator plant). Consider the transfer function

$$G(s) = \frac{1}{s^2},$$

which appears as a plant model in various applications, for example, to describe the position of a moving object that is effected by some acceleration. The plant is stabilised by the controller

$$K(s) = 12\,\frac{s+1}{s+6}.$$

Figure 3.4 shows the closed-loop step response $\bar{h}(t)$. Due to Theorem 3.4 the areas above and below the final value are equally large so that the integral of the error sums up to zero. Thus, $\bar{h}(t)$ has to overshoot the final value from which it can be concluded that the feedback loop is not externally positive. □

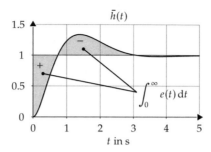

Fig. 3.4: Closed-loop step response

In order to render a closed-loop system externally positive with the output feedback depicted in Fig. 3.3, the root locus has to have branches that satisfy the necessary conditions of Lemmas 3.1–3.3. For systems with multiple eigenvalues equal to zero, these conditions are not met as proved by Theorem 3.4. Thus, in order to render a system with multiple eigenvalues equal to zero externally positive, other control structures have to be considered.

The impossibility of finding a corresponding controller results from assumption (3.13) which prohibits a controller of the form

$$K(s) = s\,\tilde{K}(s).$$

Such a controller would cancel one of the integrator eigenvalues of the plant (3.12) and, hence, invalidate the result (3.15). However, assumption (3.13) is reasonable for two reasons: On the one hand, a zero/pole cancellation with unstable poles is not robust and can to lead to closed-loop instability and, on the other hand, set-point following for ramped reference functions requires the open loop model to possess two open-loop poles equal to zero according to the internal-model principle which makes it desirable to keep the poles in the origin for this specific case.

3.3.3 Controller design for positively decomposable plants

Theorem 3.4 proved that there are classes of plants for which there does not exist any controller that leads to an externally positive feedback loop. On the other hand, there are classes for which it is easy to find an appropriate controller. This section will consider asymptotically stable and positively decomposable plants and presents a solution to the design problem by an integrating controller that also achieves set-point following for constant references based on the following observation. An integrator $G(s) = s^{-1}$ has a nonnegative impulse response as given by

$$\frac{1}{s} \multimap \sigma(t) \geq 0, \quad t \geq 0. \tag{3.16}$$

Consequently, a system of the form (3.12) with $m = 1$ has a positive decomposition if $\tilde{G}(s)$ has a positive decomposition. This result is used to prove the following theorem.

Theorem 3.5 (Synthesis for positively decomposable plants). *Consider an asymptotically stable and positively decomposable plant $G(s)$ and the integrating controller*

$$K(s) = \frac{k_I}{s}. \tag{3.17}$$

There exists an interval $\mathcal{K}_I = (0, k_I^\star]$ with the property that the closed-loop system (3.10) is externally positive

$$\bar{G}(s) \bullet\!\!-\!\!\circ \bar{g}(t) \geq 0, \quad t \geq 0$$

and asymptotically stable if the controller gain is chosen from the interval

$$k_I \in \mathcal{K}_I.$$

Moreover, the control loop has the property of set-point following for constant reference and disturbance signals.

Proof. The plant has a positive decomposition by assumption and is hence externally positive due to Lemma 3.4. The integral controller (3.17) introduces an additional pole in the origin of the complex plane. For an infinitesimal small $k_I > 0$, the closed-loop eigenvalues are equal to the open-loop eigenvalues and, hence, the closed-loop system is externally positive since it possesses a positive decomposition due to (3.12) and (3.16). For increasing controller gain k_I, the closed-loop eigenvalues move continuously along the root locus. The feedback loop stays externally positive until the integrator eigenvalue and the dominant eigenvalue of the plant meet each other in a bifurcation point. The corresponding controller gain is the upper bound $k_I^\star$. ∎

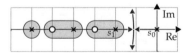

Fig. 3.5: Root locus of a positively decomposable plant with an integrating controller (3.17) and small gain k_I

The idea behind Theorem 3.5 is illustrated in Fig. 3.5 for an example plant with three poles and two zeros. The externally positive decomposition of the plant is marked by the encircled groups of poles and zeros which represent the elements (3.7) that satisfy (3.8). For $0 < k_I < k_I^\star$, the dominant pole s_1 of the plant and the additional pole s_0 of the integrator approach each other until they meet in the bifurcation point for $k_I = k_I^\star$. Thus, the additional pole of the integrator in the origin does not compromise the external positivity if the controller gain k_I is chosen sufficiently small. If k_I is increased beyond $k_I^\star$, the pole pair s_0, s_1 will split into a complex conjugate pole pair which violates the condition of Lemma 3.1 rendering the closed-loop system not externally positive.

Example 3.3 (Externally positive plant). Consider the first-order plant

$$G(s) = \frac{s - s_{01}}{s - s_1} \tag{3.18}$$

and the integrating controller (3.17). Figure 3.6 shows the root locus of the feedback system (3.10) for the values $s_{01} = -4$ and $s_1 = -3$.

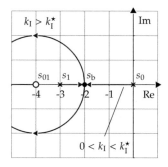

Fig. 3.6: Root locus of the combination of (3.17) and (3.18)

The two poles s_0 and s_1 approach each other for increasing controller gain k_I until they meet in the bifurcation point s_b given by

$$s_b = s_{01} \left(1 - \sqrt{1 - s_1/s_{01}}\right) = -2.$$

The corresponding controller gain equals

$$k_I^\star = s_1 - 2s_{01} - 2\sqrt{s_{01}(s_{01} - s_1)} = 1$$

at this point (cf. [89] for construction rules). The closed-loop system has two real poles located to the right of the zero s_{01} if the controller gain is chosen from the interval

$$k_I \in \mathcal{K}_I = (0, k_I^\star],$$

which renders the closed-loop impulse response nonnegative according to Theorem 3.5. □

This section showed that positively decomposable plants are positively stabilisable by an integrating controller with sufficiently small gain. Thus, the integrating controller preserves the external positivity of the plant due to the characteristic shape of the root locus that is imposed by the integrating controller. This procedure has the drawback that the plant has to be externally positive to begin with which is a stringent assumption. Furthermore, the integrating controllers slows down the dynamical behaviour of the closed-loop system in comparison to the plant which is due to the fact that the new dominant pole s_0 is closer to the origin of the complex plane than the original dominant pole s_1 of the plant.

3.4   Summary and literature notes

Theorem 3.5 and Theorem 3.4 point out different classes of systems for which there exist controllers that achieve externally positive closed-loop dynamics or not, respectively, based on the standard control configuration depicted in Fig. 3.3. If a given plant is not positively stabilisable by the standard control loop or if it is not clear how to find the controller, alternative control structures can be used like the state feedback

$$u(t) = -k^{\mathrm{T}}x(t),$$

which can place the poles of the closed-loop system arbitrarily using Ackermann's formula assuming the considered plant model is completely controllable and the state $x(t)$ is accessible by measurement or by observation.

This chapter is based on the journal contribution [6] of the author where the problem of designing externally positive feedback loop has been discussed as an open problem of control theory. This chapter serves as the theoretical fundament of Chapter 4 where the design of externally positive feedback loops is solved for vehicles.

4

Vehicle platooning

This chapter introduces the vehicle platooning problem, formalises requirements on the desired behaviour of the platoon and derives appropriate design objectives on the controlled vehicles. The main contribution of the author is presented in Section 4.3 where an adaptive cruise controller that satisfies all design objectives is synthesised for a general vehicle model. In particular, the design of externally positive feedback loops will be solved for vehicles which was an unsolved problem so far.

4.1 Introduction to adaptive cruise control

4.1.1 Description of the control problem

Adaptive cruise control (ACC) is a driver assistance system that concerns the problem of adjusting the distance in between a set of vehicles so that the inter-vehicle gaps match a desired spacing. The goal is to create a *platoon* (a stream of vehicles) of arbitrary length that is able to travel efficiently and safely along a straight road. To this end, collision avoidance is a crucial requirement that has to be ensured by each individual vehicle in the platoon.

The control problem can be formalised as follows. Consider a set of N identical vehicles that should form a platoon. The velocity of a vehicle is denoted by $v_i(t)$ and its longitudinal position by

$$s_i(t) = \int_0^t v_i(\tau)\, d\tau + s_{i0}, \quad i = 1, 2, \ldots, N.$$

The vehicles are numbered in ascending order $1, 2, \ldots, N$ beginning with the *leader*. Each *follower* $i > 1$ is equipped with sensors to measure the inter-vehicle distance

$$d_i(t) := s_{i-1}(t) - s_i(t) - d_0, \quad i = 2, 3, \ldots, N \tag{4.1}$$

to its respective predecessor $i-1$ with d_0 denoting a minimum distance which includes the vehicle length and a least required separation between two consecutive vehicles as shown in Fig. 4.1. The leader vehicle $i = 1$ is assumed to follow its piece-wise constant reference velocity $v_0(t)$ with an appropriate velocity controller.

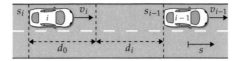

Fig. 4.1: Two consecutive vehicles

In contrast to cooperative adaptive cruise control (CACC), ACC works only with quantities that can be obtained locally without digital communication. The resulting control structure shown in Fig. 4.2 describes a predecessor following scheme which is typical for ACC platoons. The coupling of the vehicles is induced by the measurement of the distance (4.1) over the sensor network. Note that using the velocity as the coupling signal in the model is a design choice for convenience as will be shown later. The coupling of the vehicles can be equivalently described by the position.

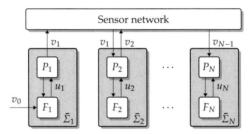

Fig. 4.2: Control structure of a platoon with ACC

This chapter will solve the ACC platooning problem in three steps based on the strategy that has been described in the introduction of this thesis.

1. The desired behaviour of the overall platoon will be specified.

2. Local properties of the controlled vehicles will be derived so that the composition of the controlled vehicles achieves the desired platoon behaviour.

3. A design procedure will be presented that solves the design problem for a general class of linear vehicle models.

The idea of using ACC instead of more powerful CACC structure is based on a safety point of view. Collision avoidance should be achieved on a fundamental level that does not require digital communication. However, the methods proposed in this chapter can be extended to CACC in order to improve performance and disturbance attenuation [88].

4.1.2 Required platoon behaviour

Since the platooning problem concerns the control of the inter-vehicle distance $d_i(t)$, the leader vehicle $i = 1$ is considered separately due to the fact that it has no predecessor. It is assumed that the leader follows its piece-wise constant nonnegative reference velocity $v_0(t) \geq 0, (t \geq 0)$ monotonically with an appropriate velocity controller F_1. That is, the velocity controller renders the controlled leader $\bar{\Sigma}_1$ asymptotically stable and externally positive with unit static reinforcement.

The followers $i = 2, 3, \ldots, N$ should satisfy the following requirements [84]:

(R1) **Asymptotic synchronisation:** In a steady state, all followers should travel with the same velocity as the leader

$$\lim_{t \to \infty} |v_i(t) - v_1(t)| = 0, \quad i = 2, 3, \ldots, N.$$

(R2) **Asymptotic time headway spacing:** The inter-vehicle distance should increase with the velocity and satisfy the requirement

$$\lim_{t \to \infty} |d_i(t) - \beta v_i(t)| = 0, \quad i = 2, 3, \ldots, N$$

with β denoting the time-headway coefficient.

(R3) **Continuous progression:** There should be no situation in which a vehicle moves backwards

$$v_i(t) \geq 0, \quad t \geq 0, \quad i = 2, 3, \ldots, N.$$

(R4) **Collision avoidance:** All inter-vehicle distances should be nonnegative

$$d_i(t) \geq 0, \quad t \geq 0, \quad i = 2, 3, \ldots, N.$$

The following section will derive properties that all controlled follower vehicles should possess individually so that they can be combined in any order to a platoon of arbitrary length that satisfies the requirements (R1)–(R4).

4.2 General design objectives

4.2.1 Model of a controlled vehicle

The requirements (R1)–(R4) should be achieved by shaping the closed-loop dynamics of the controlled vehicles with local controllers. For the sake of convenience, the notion *controlled vehicle* always refers to a follower from here on. In the frequency domain, the controlled vehicles $i = 2, 3, \ldots, N$ are described by the identical dynamics

$$\bar{\Sigma}_i : \begin{cases} V_i(s) = \bar{G}(s) \, V_{i-1}(s) \\ D_i(s) = G_d(s) \, V_{i-1}(s) \end{cases} \tag{4.2}$$

with $\bar{G}(s)$ and $G_d(s)$ denoting the closed-loop transfer functions with respect to the velocity $V_i(s) \bullet\!\!-\!\!\circ \, v_i(t)$ and the inter-vehicle distance $D_i(s) \bullet\!\!-\!\!\circ \, d_i(t)$, respectively.

The inter-vehicle distance (4.1) is equivalently described by

$$\dot{d}_i(t) = v_{i-1}(t) - v_i(t) \tag{4.3}$$

$$\circ\!\!\!\!\!\bullet$$

$$s \, D_i(s) = V_{i-1}(s) - V_i(s), \tag{4.4}$$

which yields the relation

$$G_d(s) = s^{-1} \big(1 - \bar{G}(s) \big) \tag{4.5}$$

between the transfer functions of (4.2).

Ideal vehicle dynamics

Before the design objectives are presented, this section will elaborate an ideal model of a controlled vehicle that satisfies the time-headway spacing (R2) *permanently* based on the analysis in [84]. To this end, the inter-vehicle distance has to satisfy

$$d_i(0) \stackrel{!}{=} \beta \, v_i(0) \tag{4.6}$$

initially and, furthermore, the condition

$$\dot{d}_i(t) \stackrel{!}{=} \beta \, \dot{v}_i(t), \quad t \geq 0 \tag{4.7}$$

on the derivative of the distance. The combination of (4.6) and (4.7) yields the first-order state-space model

$$\bar{\Sigma}_i : \begin{cases} \dot{v}_i(t) = -\dfrac{1}{\beta} \, v_i(t) + \dfrac{1}{\beta} \, v_{i-1}(t), \quad v_i(0) = \dfrac{d_{i0}}{\beta} \\ \begin{pmatrix} v_i(t) \\ d_i(t) \end{pmatrix} = \begin{pmatrix} 1 \\ \beta \end{pmatrix} v_i(t), \end{cases} \tag{4.8}$$

which has the transfer functions (cf. (4.2))

$$\bar{G}(s) = \frac{1}{\beta s + 1} \quad \text{and} \quad G_d(s) = \frac{\beta}{\beta s + 1}. \tag{4.9}$$

A controlled vehicle satisfies the time-headway spacing permanently if and only if it has the first-order dynamics (4.8) [84]. This observations has the following consequence: In order to achieve permanent time-headway spacing, neither the vehicle nor the controller are allowed to have any dynamical elements. This is due to the fact that the distance dynamics (4.3) already contributes with an integrator to the ideal dynamics (4.8). Any dynamical elements in the power train or the controller contributes with additional integrators to the overall dynamics and, thus, it is not possible to achieve permanent time-headway spacing with real vehicles. Nevertheless, it is possible to satisfy the requirements (R1)–(R4) if the controlled vehicles are allowed to adjust their inter-vehicle distance with a certain *delay* [86] (cf. Section 2.4.2). Then, the ideal dynamics (4.8) can be used as a benchmark to evaluate the performance of an adaptive cruise controller.

4.2.2 Asymptotic behaviour of a controlled vehicle

The control structure of a platoon with ACC in Fig. 4.2 can be rearranged to obtain the illustration in Fig. 4.3 which reveals the path structure of the platoon. Consequently, the requirements (R1) and (R2) concerning the behaviour of the platoon in a steady state can be satisfied straightforwardly by imposing the objectives

$$\bar{G}(0) = 1, \tag{4.10}$$
$$G_d(0) = \beta \tag{4.11}$$

on the static reinforcement of the controlled vehicles as shown in the following.

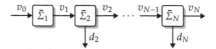

Fig. 4.3: String of coupled controlled vehicles

Consider a constant reference velocity $v_0(t) \equiv \bar{v}$ and suppose the leader already settled on the set-point due to its velocity controller, i. e.

$$v_1(t) \equiv \bar{v}, \quad t \geq 0.$$

The synchronisation as stated in (R1) requires all follower vehicles $i = 2, 3, \ldots, N$ to move with the same velocity. Given the path structure in Fig. 4.3, it follows from (4.10) that

$$v_N(\infty) = v_{N-1}(\infty) = \cdots = v_1(\infty) = \bar{v}$$

holds, which satisfies (R1). Since all vehicles move with the same velocity, the inter-vehicle distance $d_i(t)$ of each follower settles according to

$$d_i(\infty) = \beta v_{i-1}(\infty) = \beta v_i(\infty), \quad i = 2, 3, \ldots, N$$

due to the second objective (4.11). Thus, in a steady state, the inter-vehicle distances are equal to the desired time-headway spacing and (R2) is also satisfied.

Delay of a controlled vehicle. The objective (4.11) has the following interesting interpretation. The delay Δ (cf. Definition 2.2 on page 23) of a controlled vehicle can be determined with Lemma 2.1 as

$$\Delta = -\lim_{s \to 0} \frac{\mathrm{d}\bar{G}(s)}{\mathrm{d}s}. \tag{4.12}$$

Rearrangement of (4.5) yields

$$\Delta = G_\mathrm{d}(0)$$

after insertion in (4.12). Thus, the objective (4.11) requires the delay of a controlled vehicle to be equal to the desired time-headway coefficient. The reduction of the inter-vehicle gaps (e. g. to exploit slipstream) by reducing β consequently requires the controlled vehicles to have quick dynamics. Conversely, heavy vehicles with slow dynamics as trucks have to maintain a larger distance due to limitations of the control action. This limitation may be lifted by application of CACC which incorporates additional information from multiple downstream vehicles [88].

4.2.3 String stability and collision avoidance

This section will elaborate how to satisfy the requirements (R3) and (R4). In contrast to the first two requirements, (R3) and (R4) address the transient behaviour of the controlled vehicles by imposing nonnegativity of the velocity $v_i(t)$ and the distance $d_i(t)$ which is more difficult to achieve. The goal of these requirements is to ensure safety of the vehicles in the platoon and, in particular, collision avoidance. To this end, the notion of *string stability* has been introduced and discussed thoroughly in literature references concerning strings of coupled systems [132].

For a vehicle platoon in a path structure (Fig. 4.3), $\mathcal{L}_p$ string stability is defined by means of the $\mathcal{L}_p$ norm

$$\left\| v_i(t) \right\|_p := \left(\int_0^\infty \left| v_i(t) \right|^p \mathrm{d}t \right)^{1/p}$$

of the velocity as follows.

Definition 4.1 (String stability [115, 132]). A vehicle platoon is called $\mathcal{L}_p$ *string stable* if the forced zero-state response of the velocity of each controlled vehicle is bounded by its predecessor's velocity according to

$$\left\| v_i(t) \right\|_p \leq \left\| v_{i-1}(t) \right\|_p, \quad i = 2, 3, \ldots, N \tag{4.13}$$

for a reference step $v_0(t) = \bar{v}\, \sigma(t)$ of the leader.

Note that even though the notion of string stability is used to characterise a coupled system such as a vehicle platoon, the requirement (4.13) has to be satisfied by each individual controlled vehicle. Thus, it will be studied which properties the controlled vehicles have to possess to achieve (4.13).

$\mathcal{L}_2$ string stability

In the literature, the $\mathcal{L}_2$ string stability condition

$$\left\|v_i(t)\right\|_2 \leq \left\|v_{i-1}(t)\right\|_2, \quad i = 2, 3, \ldots, N \tag{4.14}$$

is often applied to achieve safety requirements on the platoon (e. g. [64, 99, 103, 114, 115, 146]). This is due to the fact that there is a simple sufficient condition on the frequency response of a controlled vehicle that results in $\mathcal{L}_2$ string stability given by the following lemma.

Lemma 4.1 (Local condition for $\mathcal{L}_2$ string stability [115]). *A vehicle platoon, that is composed of the controlled vehicles (4.2), is $\mathcal{L}_2$ string stable if the magnitude of the frequency response is bounded according to*

$$\left\|\bar{G}(s)\right\|_\infty = \sup_\omega \left|\bar{G}(j\omega)\right| \leq 1. \tag{4.15}$$

Proof. Condition (4.14) is equivalent to

$$\left\|V_i(s)\right\|_2 \leq \left\|V_{i-1}(s)\right\|_2, \quad i = 2, 3, \ldots, N$$

in the frequency domain due to Parseval's theorem [111]. With the transfer function $\bar{G}(s)$ in (4.2), the norm of the output is bounded according to

$$\left\|V_i(s)\right\|_2 \leq \left\|\bar{G}(s)\right\|_\infty \left\|V_{i-1}(s)\right\|_2 \leq \left\|V_{i-1}(s)\right\|_2.$$

Thus, $\mathcal{L}_2$ string stability can be achieved if all frequencies of the velocity $V_{i-1}(s)$ are damped which is true for transfer functions $\bar{G}(s)$ that satisfy (4.15). ∎

The condition (4.15) has been used for numerous theoretical and experimental studies to achieve $\mathcal{L}_2$ string stability. However, as proven in [84], this condition does not avoid overshooting responses of upstream vehicles which will cause collisions in long platoons with certainty as shown by the following example.

Example 4.1 ($\mathcal{L}_2$ string stable vehicles in a platoon). Consider a platoon of controlled vehicles (4.2) with the transfer function

$$\bar{G}(s) = \frac{1}{s^2 + \sqrt{2}\,s + 1}. \tag{4.16}$$

The magnitude (Fig. 4.4 (a)) of (4.16) is explicitly given by

$$\left|\bar{G}(j\omega)\right| = \frac{1}{\sqrt{1+\omega^4}},$$

which is monotonically decreasing and, hence, satisfies

$$\sup_{\omega} \left|\bar{G}(j\omega)\right| = \left|\bar{G}(0)\right| = 1. \tag{4.17}$$

Equation (4.17) shows that the transfer function (4.16) satisfies the condition (4.15). Thus, a platoon of controlled vehicles with the transfer function (4.16) is $\mathcal{L}_2$ string stable.

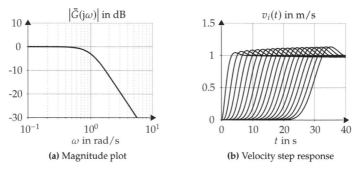

(a) Magnitude plot   **(b)** Velocity step response

Fig. 4.4: A platoon with $N = 20$ controlled vehicles (4.16)

The velocity response of a platoon with $N = 20$ controlled vehicles for the reference step $v_0(t) = \bar{v}\,\sigma(t)$ with $\bar{v} = 1\,\mathrm{m/s}$ is shown in Fig. 4.4 (b). Clearly, the $\mathcal{L}_2$ string stability of the platoon does not prevent the velocity of the vehicles to overshoot the reference value. Beyond that, the maximum value of the overshoot increases along the string. This behaviour is hazardous and will lead to a collision during an emergency braking manoeuvre as demonstrated in the following.

The distance transfer function of (4.16) is determined by (4.5) resulting in

$$G_\mathrm{d}(s) = \frac{s + \sqrt{2}}{s^2 + \sqrt{2}\,s + 1}.$$

Consider a braking manoeuvre from the reference velocity $1\,\mathrm{m/s}$ to a standstill. The response of the first follower $i = 2$ is shown in Fig. 4.5 which clearly shows that the distance $d_2(t)$ undershoots the abscissa and, hence, the requirement (R4) is not satisfied as the follower collides with its predecessor.

Discussion. This example showed that $\mathcal{L}_2$ string stability of a platoon is only necessary but not sufficient for collision avoidance due to the fact that $\mathcal{L}_2$ string stability does

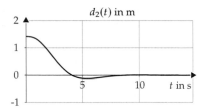

Fig. 4.5: Distance response of the second vehicle

not prohibit overshooting responses of the controlled vehicles. This is demonstrated by the transfer function (4.16) which has a pair of complex conjugate poles and, hence, an oscillating step response. □

$\mathcal{L}_\infty$ string stability

The last section demonstrated that the common $\mathcal{L}_2$ string stability condition is too weak to guarantee collision avoidance due to the possibility that the velocity responses of the controlled vehicles may overshoot and, hence, exceed the reference velocity of their respective predecessor. In order to avoid overshooting zero-state responses of upstream vehicles for a reference step of the leader, the local controllers have to guarantee that the velocity amplitude of a vehicles does not exceed the amplitude of the predecessor velocity which can be formally written as

$$\sup_{t \geq 0} |v_i(t)| \leq \sup_{t \geq 0} |v_{i-1}(t)|, \quad i = 2, 3, \ldots, N.$$

This condition is equivalent to the $\mathcal{L}_\infty$ string stability requirement

$$\left\| v_i(t) \right\|_\infty \leq \left\| v_{i-1}(t) \right\|_\infty, \quad i = 2, 3, \ldots, N. \tag{4.18}$$

The condition (4.18) has to be satisfied by the individual controlled vehicles. The following lemma will help to derive appropriate properties of the closed-loop dynamics that achieve $\mathcal{L}_\infty$ string stability of the platoon.

Lemma 4.2 (Young's convolution inequality [26]). *Let the convolution operation be denoted by $*$. Furthermore, let p, q and r denote integers that satisfy $p^{-1} + q^{-1} = r^{-1} + 1$. The norm of the convolution of two functions is bounded according to*

$$\left\| (\tilde{g} * v_{i-1})(t) \right\|_r \leq \left\| \tilde{g}(t) \right\|_p \left\| v_{i-1}(t) \right\|_q.$$

Application of Lemma 4.2 to the condition (4.18) yields the following local condition on $\mathcal{L}_\infty$ string stability.

Lemma 4.3 (Local condition for $\mathcal{L}_\infty$ string stability [37]). *Let $\bar{g}(t) \circ\!\!-\!\!\bullet \bar{G}(s)$ denote the impulse response of $\bar{\Sigma}_i$. A vehicle platoon with controlled vehicles (4.2) is $\mathcal{L}_\infty$ string stable if the impulse response satisfies*

$$\left\| \bar{g}(t) \right\|_1 = \int_0^\infty \left| \bar{g}(t) \right| dt \leq 1. \tag{4.19}$$

Proof. By choosing $r = q = \infty$ and $p = 1$, the left-hand side of (4.18) is reformulated as

$$\left\| v_i(t) \right\|_\infty = \left\| (\bar{g} * v_{i-1})(t) \right\|_\infty \leq \left\| \bar{g}(t) \right\|_1 \left\| v_{i-1}(t) \right\|_\infty$$

with Lemma 4.2. The result (4.19) follows directly as a sufficient condition. ∎

In contrast to $\mathcal{L}_2$ string stability, the condition (4.19) imposes a restriction in the time domain instead of the frequency domain. It can be trivially satisfied if the follower does not react at all, i. e. $\bar{g}(t) \equiv 0$. However, this would contradict the synchronisation requirement (R1). The following theorem establishes a connection between the synchronisation and the $\mathcal{L}_\infty$ string stability.

Theorem 4.1 (Synchronisation and $\mathcal{L}_\infty$ string stability). *Consider a controlled vehicle $\bar{\Sigma}_i$ that satisfies (4.10). It also satisfies the local condition (4.19) if and only if its impulse response $\bar{g}(t)$ is nonnegative.*

Proof. By application of the triangle inequality, the combination of condition (4.19) with design objective (4.10) yields

$$1 = \bar{G}(0) = \int_0^\infty \bar{g}(t) \, dt \leq \int_0^\infty \left| \bar{g}(t) \right| dt \leq 1. \tag{4.20}$$

Since both sides of the inequality are equal to one, it represents the equal case of the triangle inequality which is satisfied if and only if $\bar{g}(t) \geq 0, (t \geq 0)$ holds true. ∎

Theorem 4.1 requires the impulse response $\bar{g}(t)$ of the controlled vehicle $\bar{\Sigma}_i$ to be nonnegative in order to achieve $\mathcal{L}_\infty$ string stability and synchronisation of the platoon. The controlled vehicle is then said to be *externally positive* (cf. Chapter 3). Externally positive systems have a nonnegative zero-state response to a nonnegative input. This defining property can be used to satisfy the requirements (R3) and (R4) as follows.

Theorem 4.2 (Continuous progression and collision avoidance [84]). *The transient behaviour of a platoon with N controlled vehicles (4.2) satisfies the requirements (R3) and (R4) for zero initial state if the input-output behaviour of both the velocity $v_i(t)$ and the inter-vehicle distance $d_i(t)$ are externally positive with respect to the reference input $v_{i-1}(t)$:*

$$\bar{G}(s) \bullet\!\!-\!\!\circ \bar{g}(t) \geq 0, \quad t > 0$$
$$G_d(s) \bullet\!\!-\!\!\circ g_d(t) \geq 0, \quad t > 0.$$

Proof. The input-output behaviour of a controlled vehicle (4.2) is described by the convolution

$$v_i(t) = (\bar{g} * v_{i-1})(t). \tag{4.21}$$

Given the path structure of the platoon in Fig. 4.3, (4.21) can be continued as

$$v_i(t) = (\bar{g} * \bar{g} * v_{i-2})(t)$$
$$= \underbrace{(\bar{g} * \bar{g} * \cdots * \bar{g} * v_1)(t)}_{\bar{g}_{i1}(t)}.$$

Following the same reasoning, the inter-vehicle distance can be stated as

$$d_i(t) = (g_d * v_{i-1})(t) = (g_{di1} * v_1)(t).$$

If $\bar{g}(t)$ is nonnegative, $\bar{g}_{i1}(t)$ is also nonnegative for all $i = 2, 3, \dots, N$ and the same holds true for $g_{di1}(t)$ since a convolution of positive functions is again positive. Thus, if the leader's velocity $v_1(t)$ is nonnegative, then

$$v_i(t) \geq 0, \quad t \geq 0, \quad i = 2, 3, \dots, N$$
$$d_i(t) \geq 0, \quad t \geq 0, \quad i = 2, 3, \dots, N$$

hold and, hence, (R3) and (R4) are satisfied. ∎

Remark. An externally positive controlled vehicle with unit static reinforcement (4.10) is also $\mathcal{L}_2$ string stable due to Lemma 3.2, but the converse is not true. This result confirms the intuition that a stricter property than $\mathcal{L}_2$ string stability has to be applied to achieve collision avoidance.

4.2.4 Summary of the design objectives

The design objectives of the last sections guarantee that the desired behaviour of the platoon as formalised by the requirements (R1)–(R4) is achieved. In particular, the static reinforcement has to be adjusted and the controlled vehicle has to be externally positive with respect to both outputs. However, the following lemma shows that it is sufficient to just shape the impulse response $\bar{g}(t)$.

Lemma 4.4 (Externally positive vehicle [84]). *A vehicle $\bar{\Sigma}_i$ with unit static reinforcement (4.10) that is externally positive with respect to the velocity $v_i(t)$ is also externally positive with respect to the distance $d_i(t)$.*

Proof. Inverse Laplace transform of the relation (4.5) yields

$$g_d(t) = 1 - \int_0^t \bar{g}(t) \, dt. \tag{4.22}$$

The impulse response $\bar{g}(t)$ is nonnegative by assumption. Furthermore,

$$\int_0^\infty \bar{g}(t)\,\mathrm{d}t = 1$$

holds due to the unit static reinforcement (cf. (4.20)). Consequently, the indefinite integral in (4.22) is bounded according to

$$\int_0^t \bar{g}(t)\,\mathrm{d}t \in [0,1],$$

so that $g_d(t) \geq 0, (t \geq 0)$ holds true which proves the lemma. ∎

Given the considerations of the last sections, the design objectives that have to be achieved by the adaptive cruise controllers are summarised as follows [84]. Each controlled vehicle $\bar{\Sigma}_i$ has

(D1) to be asymptotically stable,

(D2) to have the static reinforcement $\bar{G}(0) = 1$,

(D3) to have the delay $G_d(0) = \beta$,

(D4) to be externally positive: $\bar{G}(s) \;\bullet\!\!-\!\!\circ\; \bar{g}(t) \geq 0, \quad t \geq 0$.

Any platoon of controlled vehicles that locally satisfy the design objectives (D1)–(D4) satisfies the requirements (R1)–(R4).

Example 4.2 (Ideal vehicle dynamics). The ideal controlled vehicle (4.8) satisfies the design objectives (D1)–(D4) as shown in the following. The closed-loop eigenvalue $-\beta^{-1}$ is stable for a positive time-headway coefficient β. The static reinforcements with respect to the velocity $v_i(t)$ and the distance $d_i(t)$ can be evaluated straightforwardly using (4.9). The impulse response of (4.8) is

$$\bar{g}(t) = \beta^{-1} e^{-\beta^{-1} t},$$

which is nonnegative due to the scalar exponential function (Fig. 4.6 (a)). Consequently, a platoon of ideal controlled vehicles (4.8) satisfies the requirements (R1)–(R4) and, in particular, guarantees collision avoidance in a platoon of arbitrary length N.

The velocity responses in Fig. 4.6 (b) of a platoon that is composed of $N = 20$ ideal vehicles with $\beta = 1\,\mathrm{s}$ monotonically follow the reference velocity $\bar{v} = 1\,\mathrm{m/s}$ due to the externally positive dynamics. □

The design objectives fully characterise the ideal dynamics (4.8) but not any linear system in general and, hence, there is freedom left in shaping the closed-loop dynamics of the controlled vehicles. Thus, the controlled vehicles do not have to be identical to satisfy

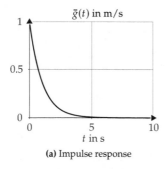

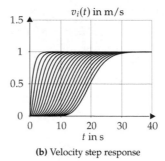

Fig. 4.6: A platoon with $N = 20$ ideal controlled vehicles (4.8)

the requirements (R1)–(R4). The platoons with identical dynamics have been considered solely for convenience.

A key element of the design objectives is to achieve externally positive closed-loop dynamics. As elaborated in Chapter 3, the design of externally positive feedback loops is a hard task and not solved in general [6]. The limiting factor is the fact that the impulse response of a feedback loop is hardly determinable as a function of the controller parameters. As a main result of this thesis, the following section will solve the ACC design problem for a general class of linear vehicle models using a decomposition of the closed-loop impulse response.

4.3 Design of adaptive cruise controllers

4.3.1 Way of solution

The goal of this section is to synthesize an adaptive cruise controller that achieves the design objectives (D1)–(D4). To this end, a control structure has to be specified that provides sufficient degrees of freedom and, then, the controller parameters have to be determined that solve the design problem.

In particular, this section provides a method to design externally positive feedback loops with a state feedback approach. The external positivity will be achieved with the factorisation approach summarised by Theorem 3.3 in Chapter 3. The idea is to shape the closed-loop transfer function $\bar{G}(s)$ so that it admits the decomposition

$$\bar{G}(s) = \kappa \prod_{i=1}^{\tilde{n}} \bar{G}_i(s), \quad \kappa > 0 \tag{4.23}$$

into a series of subsystems $\bar{G}_i(s), (i = 1, 2, \ldots, \tilde{n})$ that satisfy

$$\bar{G}_i(s) \bullet\!\!-\!\!\circ \bar{g}_i(t) \geq 0, \quad t \geq 0, \quad i = 1, 2, \ldots, \tilde{n}. \tag{4.24}$$

The transfer function $\bar{G}(s)$ then has a *positive decomposition* and is externally positive due to Lemma 3.4: $\bar{G}(s) \bullet\!\!-\!\!\circ \bar{g}(t) \geq 0, (t \geq 0)$.

Before the control structure and the parametrisation will be presented, the next section introduces the class of considered vehicle models.

4.3.2 Vehicle model

The velocity $v_i(t)$ of vehicle i is determined by the strictly proper n-th order state-space model

$$\Sigma_i : \begin{cases} \dot{x}_i(t) = A\,x_i(t) + b\,u_i(t), & x_i(0) = x_{i0} \\ v_i(t) = c^{\mathrm{T}} x_i(t), \end{cases} \tag{4.25}$$

which will be referred to as the *vehicle model*. Its state $x_i(t) \in \mathbb{R}^n$ may represent (but is not limited to) physical quantities of the power train, e. g. acceleration, jerk or torque. The scalar input $u_i(t)$ is, for example, the commanded acceleration of the power train. For $x_{i0} = 0$, the input-output behaviour of Σ_i is equivalently described by

$$\Sigma_i : \ V_i(s) = G(s)\,U_i(s)$$

with the transfer function

$$G(s) = c^{\mathrm{T}}(sI - A)^{-1}b. \tag{4.26}$$

Note that the model parameters (A, b, c^{T}) do not have to be identical for each vehicle. Since the design procedure that will be elaborated in the following considers one single vehicle, an individual index i is omitted. The following assumptions are imposed on Σ_i.

Assumption 4.1 (Controllability). The state-space model (4.25) is a minimal realisation of $G(s)$, i. e. Σ_i is controllable and observable. Consequently, the controllability matrix

$$S := \begin{pmatrix} b & Ab & \ldots & A^{n-1}b \end{pmatrix} \tag{4.27}$$

is regular.

Assumption 4.2 (No zeros). Σ_i has no transmission zeros, i. e. there does not exist any frequency μ that satisfies $G(\mu) = 0$.

Due to Assumption 4.2, the transfer function (4.26) is of the form

$$G(s) = \frac{1}{s^p} \cdot \frac{b_0}{\sum_{j=0}^{n-p} a_j s^j}, \quad 0 \leq p \leq n,$$

which represents a series interconnection of p integrators and an $(n-p)$-th order lag system. This assumption simplifies the design process in several ways as will be shown in the following sections. Apart from these assumptions, the coefficients b_0 and a_j are arbitrary. The models

$$G(s) = \frac{1}{s(\tau s + 1)} \quad \text{and} \quad G(s) = \frac{1}{ms + c} \tag{4.28}$$

are often encountered in the literature, e. g. in [114] and [102] which are much more specific than the vehicle model used here. In the first model, the control input $u_i(t)$ influences the acceleration through a lag with the time constant τ and in the second model, the control input represents an accelerating force that works on the lossy vehicle with the mass m and the friction constant c. Both models comply with Assumption 4.2.

Note that there are studies that consider nonlinear vehicle dynamics, e. g. [140] where a velocity dependent friction $\zeta(v_i(t))$ is applied to model the longitudinal dynamics:

$$m \, \dot{v}_i(t) = \zeta\big(v_i(t)\big) \, v_i(t) + u_i(t). \tag{4.29}$$

To apply the methods that will be proposed in this chapter to achieve external positivity, it is possible to linearise the nonlinear dynamics (4.29) to achieve a linear model of the form (4.28) [87].

Coupled vehicle model. The vehicle model (4.25) describes the velocity dynamics of a single vehicle without the influence of other vehicles. By adding the distance dynamics (4.3), the coupled vehicle model

$$P_l : \begin{cases} \begin{pmatrix} \dot{x}_i(t) \\ \dot{d}_i(t) \end{pmatrix} = \begin{pmatrix} A & 0 \\ -c^{\mathsf{T}} & 0 \end{pmatrix} \begin{pmatrix} x_i(t) \\ d_i(t) \end{pmatrix} + \begin{pmatrix} b \\ 0 \end{pmatrix} u_i(t) + \begin{pmatrix} 0 \\ 1 \end{pmatrix} v_{i-1}(t) \\ \begin{pmatrix} v_i(t) \\ d_i(t) \end{pmatrix} = \begin{pmatrix} c^{\mathsf{T}} & 0 \\ 0^{\mathsf{T}} & 1 \end{pmatrix} \begin{pmatrix} x_i(t) \\ d_i(t) \end{pmatrix}, \quad \begin{pmatrix} x_i(0) \\ d_i(0) \end{pmatrix} = \begin{pmatrix} x_{i0} \\ d_{i0} \end{pmatrix} \end{cases} \tag{4.30}$$

is obtained (Fig. 4.7) which is coupled with its predecessor $i-1$ due to the sensor network (cf. Fig. 4.2).

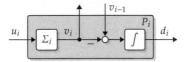

Fig. 4.7: Coupled vehicle model ($i = 2, 3, \ldots, N$)

4.3.3 Extended state feedback control structure

In order to achieve the required time headway spacing (R2), the controller state

$$\dot{z}_i(t) = \beta\, v_i(t) - d_i(t), \quad z_i(0) = z_{i0} \tag{4.31}$$

is introduced. The right-hand side of (4.31) represents the time-headway error which should be equal to zero in the steady state. The combination of the coupled vehicle model (4.30) and the controller state (4.31) yields the open-loop state equation

$$\begin{pmatrix} \dot{x}_i(t) \\ \dot{d}_i(t) \\ \dot{z}_i(t) \end{pmatrix} = \underbrace{\begin{pmatrix} A & 0 & 0 \\ -c^{\mathrm{T}} & 0 & 0 \\ \beta c^{\mathrm{T}} & -1 & 0 \end{pmatrix}}_{A_0} \begin{pmatrix} x_i(t) \\ d_i(t) \\ z_i(t) \end{pmatrix} + \underbrace{\begin{pmatrix} b \\ 0 \\ 0 \end{pmatrix}}_{b_0} u_i(t) + \underbrace{\begin{pmatrix} 0 \\ 1 \\ 0 \end{pmatrix}}_{e} v_{i-1}(t), \tag{4.32}$$

which has two eigenvalues equal to zero as the structure of the matrix A_0 reveals. As shown by Theorem 3.4, the standard control loop with the output feedback in Fig. 3.3 cannot render a system with two eigenvalues equal to zero externally positive. Thus, the feedback

$$u_i(t) = -k^{\mathrm{T}} \begin{pmatrix} x_i(t) \\ d_i(t) \\ z_i(t) \end{pmatrix} = -\begin{pmatrix} k_{\mathrm{x}}^{\mathrm{T}} & k_{\mathrm{d}} & k_{\mathrm{z}} \end{pmatrix} \begin{pmatrix} x_i(t) \\ d_i(t) \\ z_i(t) \end{pmatrix} \tag{4.33}$$

of the state $x_i(t)$ of the vehicle (4.25), the inter-vehicle distance $d_i(t)$ and the controller state $z_i(t)$ is applied. In contrast to an output feedback, the control law (4.33) allows for more freedom in shaping the closed-loop dynamics. The resulting control loop is illustrated in Fig. 4.8.

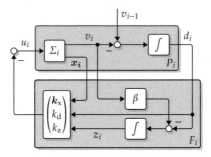

Fig. 4.8: Structure of the feedback loop

The combination of (4.32) and (4.33) leads to the model

$$
\bar{\Sigma}_i : \begin{cases}
\begin{pmatrix} \dot{\boldsymbol{x}}_i(t) \\ \dot{d}_i(t) \\ \dot{z}_i(t) \end{pmatrix} = \bar{A} \begin{pmatrix} \boldsymbol{x}_i(t) \\ d_i(t) \\ z_i(t) \end{pmatrix} + \boldsymbol{e}\, v_{i-1}(t), \quad \begin{pmatrix} \boldsymbol{x}_i(0) \\ d_i(0) \\ z_i(0) \end{pmatrix} = \begin{pmatrix} \boldsymbol{x}_{i0} \\ d_{i0} \\ z_{i0} \end{pmatrix} \\[4ex]
v_i(t) = \underbrace{\begin{pmatrix} \boldsymbol{c}^{\mathrm{T}} & 0 & 0 \end{pmatrix}}_{\bar{\boldsymbol{c}}^{\mathrm{T}}} \begin{pmatrix} \boldsymbol{x}_i(t) \\ d_i(t) \\ z_i(t) \end{pmatrix} \\[4ex]
d_i(t) = \underbrace{\begin{pmatrix} \boldsymbol{0}^{\mathrm{T}} & 1 & 0 \end{pmatrix}}_{\boldsymbol{c}_{\mathrm{d}}^{\mathrm{T}}} \begin{pmatrix} \boldsymbol{x}_i(t) \\ d_i(t) \\ z_i(t) \end{pmatrix}
\end{cases} \tag{4.34}
$$

of the controlled vehicle with $\boldsymbol{e}$ defined in (4.32) and

$$
\bar{A} = A_0 - b_0 \boldsymbol{k}^{\mathrm{T}} = \begin{pmatrix} A - b\boldsymbol{k}_{\mathrm{x}}^{\mathrm{T}} & -b k_{\mathrm{d}} & -b k_z \\ -\boldsymbol{c}^{\mathrm{T}} & 0 & 0 \\ \beta \boldsymbol{c}^{\mathrm{T}} & -1 & 0 \end{pmatrix}, \quad \dim \bar{A} = n + 2.
$$

Implementation of the controller. The local feedback loop is illustrated in Fig. 4.8. The controller works with local quantities and without digital communication because the distance $d_i(t)$ is measured with appropriate sensors. Thus, in order to implement the proposed control structure, the state $\boldsymbol{x}_i(t)$ has to be known. For that, the state either has to be measured directly or it has to be reconstructed with an observer or estimator. In the latter case, there are well studied methods to observe the state with a Luenberger observer (cf. [79]) or a Kalman filter which has been applied to estimate the state of a vehicle in [63]. The measurement of the distance $d_i(t)$ is assumed to be instantaneous. However, if the controller receives the distance information with the delay t_Δ, it has to be made sure that the time-headway coefficient β exceeds the delay

$$
\beta > t_\Delta
$$

to apply the method proposed in this paper. Due to the path structure given by (4.3), the stability of the overall platoon is not jeopardised. A thorough discussion on this topic is given in [86].

Closed-loop transmission zeros

Although the vehicle model (4.25) has no transmission zeros by Assumption 4.2, the controlled vehicle (4.34) has a transmission zero whose location depends on the controller parameters as shown by the following lemma.

Lemma 4.5 (Transmission zero). *The controlled vehicle* (4.34) *has a single closed-loop transmission zero* $\bar{\mu}$ *at*

$$\bar{\mu} = \frac{k_z}{k_d}. \tag{4.35}$$

Proof. Let $|\cdot|$ denote the determinant of a matrix. The zeros of the controlled vehicle (4.34) are determined by the roots of the determinant of the Rosenbrock system matrix as

$$
|P(\bar{\mu})| = \begin{vmatrix} \bar{\mu}I - \bar{A} & -e \\ \bar{c}^{\mathrm{T}} & 0 \end{vmatrix} = \begin{vmatrix} \bar{\mu}I - A + bk_x^{\mathrm{T}} & bk_d & bk_z & 0 \\ c^{\mathrm{T}} & \bar{\mu} & 0 & -1 \\ -\beta c^{\mathrm{T}} & 1 & \bar{\mu} & 0 \\ c^{\mathrm{T}} & 0 & 0 & 0 \end{vmatrix}
$$

$$
= -1 \begin{vmatrix} \bar{\mu}I - A & 0 & b \\ -\beta c^{\mathrm{T}} & 1 & 0 \\ c^{\mathrm{T}} & 0 & 0 \end{vmatrix} \cdot \begin{vmatrix} I & 0 & 0 \\ 0^{\mathrm{T}} & 1 & \bar{\mu} \\ -k_x^{\mathrm{T}} & -k_d & -k_z \end{vmatrix}
$$

$$
= -1 \underbrace{\begin{vmatrix} \bar{\mu}I - A & -b \\ c^{\mathrm{T}} & 0 \end{vmatrix}}_{\neq 0, \; \forall \bar{\mu}} \cdot \begin{vmatrix} 1 & \bar{\mu} \\ -k_d & -k_z \end{vmatrix} = 0.
$$

The first factor is nonzero for any $\bar{\mu}$ due to Assumptions 4.1–4.2 and the second factor is singular for the zero (4.35). ∎

The closed-loop transmission zero is important because of the necessary condition on external positivity stated by Lemma 3.3 which claims that the zeros of an externally positive system have to be located to the left of the dominant eigenvalue. Based on this result, the choice of the feedback gain k has to ensure that the transmission zero $\bar{\mu}$ is located sufficiently far in the left half plane to satisfy this necessary condition for external positivity. This requirement leads to a fundamental problem: It is well known in multivariable control theory that the closed-loop eigenvalues of a completely controllable system can be placed arbitrarily in the complex plane by a feedback of the state through one single control input [90]. However, in order to place $\bar{\mu}$ arbitrarily, additional degrees of freedom are necessary in general. Thus, the choice of the closed-loop eigenvalues has to consider the resulting location of $\bar{\mu}$ so that the design problem is rendered solvable. Beforehand, the asymptotic behaviour of the control loop in Fig. 4.8 will be analysed.

Asymptotic behaviour of the control loop

The ACC feedback loop in Fig. 4.8 satisfies the design objectives (D2) and (D3) concerning the static behaviour as shown in the following. The feedback loop has two integrators with the respective inputs $v_{i-1}(t) - v_i(t)$ and $\beta v_i(t) - d_i(t)$. Thus, if the feedback gain k stabilises the closed-loop system (4.34), the inputs of both integrators have to vanish asymptotically, i.e. the requirements (R1) and (R2) are satisfied. This observation is independent of the controller parameters since the feedback loop satisfies the desired static behaviour structurally as shown by the following lemma.

Lemma 4.6 (Static behaviour of the controlled vehicle). *Consider the controlled vehicle (4.34). The static reinforcement of the outputs $v_i(t)$ and $d_i(t)$ with respect to the input $v_{i-1}(t)$ is $\bar{G}(0) = 1$ and $G_d(0) = \beta$, respectively.*

Proof. The static reinforcements can be determined with the elements of the state-space model (4.34) according to

$$\bar{G}(0) = -\bar{c}^T \bar{A}^{-1} e, \tag{4.36}$$

$$G_d(0) = -c_d^T \bar{A}^{-1} e. \tag{4.37}$$

In order to evaluate these equations, the inverse of the closed-loop system matrix $\bar{A}$ is obtained with Gauss–Jordan elimination which results in

$$\bar{A}^{-1} = \begin{pmatrix} (A - bk_x^T)^{-1} \left(I - bc^T \dfrac{(A - bk_x^T)^{-1}}{c^T(A - bk_x^T)^{-1}b} \right) & -\dfrac{(A - bk_x^T)^{-1}b}{c^T(A - bk_x^T)^{-1}b} & 0 \\[3mm] 0^T & -\beta & -1 \\[3mm] -\dfrac{c^T(A - bk_x^T)^{-1}}{c^T(A - bk_x^T)^{-1}bk_z} & \beta\dfrac{k_d}{k_z} - \dfrac{1}{c^T(A - bk_x^T)^{-1}bk_z} & \dfrac{k_d}{k_z} \end{pmatrix}. \tag{4.38}$$

Straightforward evaluation of (4.36) and (4.37) with (4.38) and the elements of (4.34) completes the proof. ∎

Lemma 4.6 shows that the feedback loop in Fig. 4.8 satisfies the design objectives (D2) and (D3) independently of the feedback gain k. This result simplifies the design problem as follows. For the determination of the feedback gain k, the static behaviour of the feedback loop does not have to be considered. Thus, the design procedure that will be proposed in the next section can focus on the stability and external positivity of the controlled vehicle. This important result is summarised in the following theorem.

Theorem 4.3 (ACC with state feedback). *The feedback loop in Fig. 4.8 satisfies the design objectives (D1)–(D4) if and only if the feedback gain k renders the controlled vehicle (4.34) asymptotically stable and externally positive.*

4.3.4 Determination of the feedback gain

The application of the state feedback (4.33) allows for using pole assignment techniques [90]. To this end, the controlled vehicle (4.34) will be analysed in order to find a set of closed-loop eigenvalues $\bar{\lambda}_i$, $(i = 1, 2, \ldots, n+2)$ that will be installed by pole assignment so that the closed-loop transfer function $\bar{G}(s)$ has the positive decomposition (4.23)–(4.24).

By using the pole assignment approach, the feedback gain k is determined by Ackermann's formula (cf. [18])

$$k^\mathrm{T} = \begin{pmatrix} k_x^\mathrm{T} & k_\mathrm{d} & k_z \end{pmatrix} = s_{R0}^\mathrm{T} \bar{P}(A_0) \tag{4.39}$$

with the characteristic polynomial of the controlled vehicle

$$\bar{P}(A_0) = \bar{a}_0 I + \bar{a}_1 A_0 + \cdots + \bar{a}_{n+1} A_0^{n+1} + A_0^{n+2}, \tag{4.40}$$

which is evaluated for the open-loop system matrix A_0 defined in (4.32), and the vector

$$s_{R0}^\mathrm{T} = \begin{pmatrix} 0 & \cdots & 0 & 1 \end{pmatrix} S_0^{-1}, \tag{4.41}$$

where the matrix

$$S_0 = \begin{pmatrix} b_0 & A_0 b_0 & \ldots & A_0^{n+1} b_0 \end{pmatrix} \tag{4.42}$$

denotes the controllability matrix of the open-loop model (4.32).

Instead of seeking the feedback gain k as a whole, an explicit representation of the parameters k_d and k_z that determine the closed-loop transmission zero $\bar{\mu}$ in (4.35) is derived by decomposing Ackermann's formula as follows.

Lemma 4.7 (Feedback gain). *The feedback parameters k_d and k_z in (4.39) are given by*

$$k_\mathrm{d} = \frac{\bar{a}_0 \beta - \bar{a}_1}{c^\mathrm{T} A^{n-1} b} \quad and \quad k_z = \frac{\bar{a}_0}{c^\mathrm{T} A^{n-1} b}. \tag{4.43}$$

Proof. The lemma is proved in two steps. Both the vector s_{R0}^T and the characteristic polynomial $\bar{P}(A_0)$ are decomposed in the following to obtain the result (4.43).

Assumption 4.2 will be applied to decompose the open loop controllability matrix

$$S_0 = \begin{pmatrix} b & Ab & A^2 b & \cdots & A^{n-1} b & \vdots & A^n b & A^{n+1} b \\ \hline 0 & -c^\mathrm{T} b & -c^\mathrm{T} Ab & \cdots & -c^\mathrm{T} A^{n-2} b & \vdots & -c^\mathrm{T} A^{n-1} b & -c^\mathrm{T} A^n b \\ 0 & \beta c^\mathrm{T} b & c^\mathrm{T} (I + \beta A) b & \cdots & c^\mathrm{T} (I + \beta A) A^{n-3} b & \vdots & c^\mathrm{T} (I + \beta A) A^{n-2} b & c^\mathrm{T} (I + \beta A) A^{n-1} b \end{pmatrix} \tag{4.44}$$

as follows. The relative degree r of a system without zeros is equal to its order: $r = n$. As a consequence, the Markov parameters of the vehicle model Σ_i satisfy the properties (cf. [89])

$$c^\mathrm{T} A^i b = 0, \quad i = 0, 1, \ldots, n-2, \tag{4.45}$$

$$c^\mathrm{T} A^{n-1} b \neq 0. \tag{4.46}$$

Application of (4.45) and (4.46) to S_0 yields a decomposition in block triangular form as

$$S_0 = \begin{pmatrix} S & R \\ O & Q \end{pmatrix} \tag{4.47}$$

with the controllability matrix S of the vehicle model (4.25) defined in (4.27) and

$$R = \begin{pmatrix} A^n b & A^{n+1} b \end{pmatrix} \in \mathbb{R}^{n \times 2}$$

$$Q = \begin{pmatrix} -c^T A^{n-1} b & -c^T A^n b \\ \beta c^T A^{n-1} b & c^T (I + \beta A) A^{n-1} b \end{pmatrix} \in \mathbb{R}^{2 \times 2}.$$

The $(2 \times n)$-zero-matrix O in the bottom left of (4.47) results from the fact that the first n elements in the last two rows of (4.44) are equal to zero due to (4.45).

With the decomposition of S_0, the determination of the last row of its inverse in (4.41) is equivalently stated as

$$s_{R0}^T S_0 = \begin{pmatrix} 0 & \cdots & 0 & 1 \end{pmatrix}$$

$$\Longleftrightarrow \begin{pmatrix} s_R^T & q^T \end{pmatrix} \begin{pmatrix} S & R \\ O & Q \end{pmatrix} = \begin{pmatrix} 0 & \cdots & 0 & 1 \end{pmatrix} \tag{4.48}$$

by introduction of

$$s_{R0}^T = \begin{pmatrix} s_R^T & q^T \end{pmatrix}. \tag{4.49}$$

Equation (4.48) is then decomposed into

$$s_R^T S = 0^T$$

$$s_R^T R + q^T Q = \begin{pmatrix} 0 & 1 \end{pmatrix}.$$

The first equation holds true if and only if

$$s_R^T = 0^T \tag{4.50}$$

since S is regular due to Assumption 4.1. Hence, the second equation simplifies to

$$q^T Q = \begin{pmatrix} 0 & 1 \end{pmatrix},$$

which can be solved by straightforward inversion of Q:

$$q^T = \begin{pmatrix} 0 & 1 \end{pmatrix} Q^{-1} = \frac{1}{c^T A^{n-1} b} \begin{pmatrix} \beta & 1 \end{pmatrix}. \tag{4.51}$$

In summary, the first factor of Ackermann's formula (4.39) is

$$s_{R0}^T = \begin{pmatrix} 0 & \cdots & 0 & \dfrac{\beta}{c^T A^{n-1} b} & \dfrac{1}{c^T A^{n-1} b} \end{pmatrix}.$$

In order to evaluate the characteristic polynomial (4.40) of the controlled vehicle, the m-th power of A_0 in (4.32)

$$A_0^m = \left(\begin{array}{cc|cc} A^m & & 0 & 0 \\ \hline -c^{\mathsf{T}} A^{m-1} & & 0 & 0 \\ c^{\mathsf{T}} A^{m-2}(\beta A + I) & & 0 & 0 \end{array} \right), \quad m > 1 \tag{4.52}$$

has to be examined. Equation (4.52) shows that the last two columns are equal to zero for exponents $m > 1$. Thus, the characteristic polynomial (4.40) has the block triangular form

$$\bar{P}(A_0) = \begin{pmatrix} \bar{P}(A) & O \\ * & T \end{pmatrix} \tag{4.53}$$

with $*$ denoting a nonzero $(2 \times n)$-matrix and

$$T = \begin{pmatrix} \bar{a}_0 & 0 \\ -\bar{a}_1 & \bar{a}_0 \end{pmatrix}, \tag{4.54}$$

which results from the bottom right (2×2)-matrix of the first two summands of $\bar{P}(A_0)$. Applying (4.49), (4.50) and (4.53) to Ackermann's formula (4.39) yields

$$\begin{pmatrix} k_x^{\mathsf{T}} & k_d & k_z \end{pmatrix} = \begin{pmatrix} 0^{\mathsf{T}} & q^{\mathsf{T}} \end{pmatrix} \begin{pmatrix} \bar{P}(A) & O \\ * & T \end{pmatrix}.$$

Consequently, the parameters k_d and k_z that determine the closed-loop transmission zero (4.35) are given by

$$\begin{pmatrix} k_d & k_z \end{pmatrix} = q^{\mathsf{T}} T$$

or, explicitly, by

$$k_d = \frac{\bar{a}_0 \beta - \bar{a}_1}{c^{\mathsf{T}} A^{n-1} b} \quad \text{and} \quad k_z = \frac{\bar{a}_0}{c^{\mathsf{T}} A^{n-1} b}$$

using the results (4.51) and (4.54) which completes the proof. ∎

Lemma 4.7 solves Ackermann's formula (4.39) explicitly for the parameters k_d and k_z. This is an important result due to the fact that these parameters determine the closed-loop zero (4.35). Note that it is not possible to determine k_x explicitly without further knowledge of the system matrix A of the vehicle model (4.25). However, this limitation is not a problem concerning the design objectives as shown in the following section.

4.3.5 Determination of eigenvalues that achieve external positivity

The goal of this section is to determine a set of closed-loop eigenvalues $\bar{\lambda}_i, (i = 1, 2, \ldots, n+2)$ that renders the controlled vehicle asymptotically stable and externally positive. Once these eigenvalues are determined, they can be installed by pole assignment using Ackermann's formula which solves the design problem. To this end, the closed-loop transmission zero will be analysed in the following.

Relation between the closed-loop zero and the eigenvalues

With the result (4.43), the relation

$$\bar{\mu} = \frac{k_z}{k_d} = \frac{1}{\beta - \bar{a}_1/\bar{a}_0} \tag{4.55}$$

between the closed-loop zero $\bar{\mu}$ and the closed-loop eigenvalues $\bar{\lambda}_i$ is obtained. In order to establish a connection between $\bar{\mu}$ and $\bar{\lambda}_i$, consider the factorised form of the closed-loop characteristic polynomial

$$\bar{P}(\lambda) = \sum_{i=0}^{n+2} \bar{a}_i \lambda^i = \prod_{i=1}^{n+2} (\lambda - \bar{\lambda}_i) \tag{4.56}$$

with $\bar{a}_{n+2} = 1$. For a given set of eigenvalues $\bar{\lambda}_i$, the coefficients $\bar{a}_i$ of the polynomial are determined by Vieta's formulas [142]. The first two coefficients are given by

$$\bar{a}_0 = (-1)^{n+2} \prod_{i=1}^{n+2} \bar{\lambda}_i, \tag{4.57}$$

$$\bar{a}_1 = (-1)^{n+1} \sum_{j=1}^{n+2} \prod_{\substack{i=1 \\ i \neq j}}^{n+2} \bar{\lambda}_i = (-1)^{n+1} \prod_{i=1}^{n+2} \bar{\lambda}_i \sum_{j=1}^{n+2} \frac{1}{\bar{\lambda}_j}, \tag{4.58}$$

which are sufficient to determine the feedback parameters (4.43) and the zero (4.55). The relation between the closed-loop zero and the eigenvalues is then given by the following lemma.

Lemma 4.8 (Determination of the transmission zero). *The eigenvalues $\bar{\lambda}_i$, ($i = 1, 2, \ldots, n+2$) of the controlled vehicle (4.34) determine the zero $\bar{\mu}$ according to*

$$\bar{\mu} = \frac{1}{\beta + \sum_{i=1}^{n+2} \bar{\lambda}_i^{-1}}. \tag{4.59}$$

Proof. The expressions (4.57) and (4.58) yield the fraction

$$\frac{\bar{a}_1}{\bar{a}_0} = -\sum_{i=1}^{n+2} \frac{1}{\bar{\lambda}_i},$$

which results in (4.59) after insertion in (4.55). ∎

Conditions for closed-loop external positivity

This section presents conditions on the choice of the closed-loop eigenvalues $\bar{\lambda}_i$ so that the controlled vehicle (4.34) is asymptotically stable and externally positive for a given β. If the closed-loop eigenvalues are chosen to satisfy the order

$$\text{Re}\{\bar{\lambda}_i\} < \text{Re}\{\bar{\lambda}_1\} < 0, \quad i = 2, 3, \ldots, n+2 \tag{4.60}$$

asymptotic stability is guaranteed with $\bar{\lambda}_1$ denoting the dominant eigenvalue of the controlled vehicle (cf. Definition 3.2). Lemma 4.8 shows that the choice of the time-headway coefficient β and the closed-loop eigenvalues $\bar{\lambda}_i$ determine the transmission zero $\bar{\mu}$. Due to Lemma 3.3, this choice has to make sure that $\bar{\mu}$ is located to the left of the dominant eigenvalue. With the relation (4.59), this is achieved if and only if the condition

$$\beta + \sum_{i=2}^{n+2} \frac{1}{\bar{\lambda}_i} > 0 \tag{4.61}$$

holds true. By satisfying (4.60) and (4.61) with an appropriate choice of the closed-loop eigenvalues for a given β, the controlled vehicle (4.34) is asymptotically stable and externally positive as shown by the following theorem.

Theorem 4.4 (Eigenvalues that achieve asymptotic stability and external positivity). *The controlled vehicle (4.34) satisfies the design objectives (D1)–(D4) if its eigenvalues $\bar{\lambda}_i, (i = 1, 2, \ldots, n+2)$ are chosen so as to satisfy the requirements*

$$\text{Re}\{\bar{\lambda}_i\} < \text{Re}\{\bar{\lambda}_1\} < 0, \quad i = 2, 3, \ldots, n+2 \tag{4.62}$$

$$\text{Im}\{\bar{\lambda}_i\} = 0, \quad i = 1, 2, \ldots, n+2 \tag{4.63}$$

$$\beta + \sum_{i=2}^{n+2} \frac{1}{\bar{\lambda}_i} > 0. \tag{4.64}$$

Proof. Equation (4.62) implies asymptotic stability which satisfies (D1)–(D3) due to Lemma 4.6. The transfer function $\bar{G}(s)$ of the controlled vehicle is constructed as

$$\bar{G}(s) = \kappa \frac{s - \bar{\mu}}{\prod_{i=1}^{n+2}(s - \bar{\lambda}_i)}$$

with κ denoting the scalar factor determined by

$$\kappa = -\bar{\mu}^{-1} \prod_{i=1}^{n+2}(-\bar{\lambda}_i)$$

due to $\bar{G}(0) = 1$. Thus, $\bar{G}(s)$ admits the positive decomposition

$$\bar{G}(s) = \prod_{i=1}^{n+2} \bar{G}_i(s) = \underbrace{\frac{\bar{\lambda}_1}{\bar{\mu}} \frac{s - \bar{\mu}}{s - \bar{\lambda}_1}}_{\bar{G}_1(s)} \prod_{i=2}^{n+2} \frac{-\bar{\lambda}_i}{s - \bar{\lambda}_i}$$

with the correspondences

$$\bar{G}_1(s) \bullet\!\!-\!\!\circ \bar{g}_1(t) = \frac{\bar{\lambda}_1}{\bar{\mu}} (\bar{\lambda}_1 - \bar{\mu}) e^{\bar{\lambda}_1 t} \geq 0, \quad t \geq 0,$$

$$\bar{G}_i(s) \bullet\!\!-\!\!\circ \bar{g}_i(t) = -\bar{\lambda}_i e^{\bar{\lambda}_i t} \geq 0, \quad t \geq 0, \quad i = 2, 3, \ldots, n + 2,$$

which are nonnegative due to (4.62) and (4.63) and, hence, (D4) is also satisfied. ∎

Remark. Theorem 4.4 can be relaxed to allow complex conjugate nondominant eigenvalues, i.e. condition (4.63) can be replaced by $\text{Im}\{\bar{\lambda}_1\} = 0$ since it is only necessary that the dominant eigenvalue is real.

Theorem 4.4 provides clear conditions on the choice of the closed-loop eigenvalues so that the controlled vehicle satisfies the design objectives (D1)–(D4). However, the conditions do not fully characterise the controlled vehicle (4.34) since they allow for some freedom in choosing the closed-loop eigenvalues. Furthermore, it is not clear straightforwardly how to satisfy (4.64). The following section will address this observation and propose an explicit procedure to find a set of closed-loop eigenvalues that satisfy the conditions of Theorem 4.4.

Closed-loop eigenvalues that achieve external positivity

In order to find a set of eigenvalues that satisfies the conditions (4.62)–(4.64), the following simplifications will be imposed. The eigenvalues are supposed to be real and ordered according to

$$\bar{\lambda}_{n+2} \leq \bar{\lambda}_{n+1} \leq \ldots \leq \bar{\lambda}_2 < \bar{\lambda}_1 < 0 \tag{4.65}$$

and the closed-loop zero should cancel with the smallest closed-loop eigenvalue:

$$\bar{\mu} \overset{!}{=} \bar{\lambda}_{n+2}. \tag{4.66}$$

With (4.66), the relation (4.59) is equivalently stated as

$$\bar{\lambda}_{n+2} \left(\beta + \sum_{i=1}^{n+1} \frac{1}{\bar{\lambda}_i} \right) = 0$$

and since $\bar{\lambda}_{n+2} < 0$ due to (4.65), it follows that

$$\beta + \sum_{i=1}^{n+1} \frac{1}{\bar{\lambda}_i} = 0. \tag{4.67}$$

The condition (4.67) is an alternative representation of the condition (4.64) with the difference that the location of the zero $\bar{\mu}$ is determined by the choice of an eigenvalue. The positive decomposition of $\bar{G}(s)$ is not jeopardised due to $\bar{\lambda}_{n+2} < \bar{\lambda}_1$ and the following observation. By adding and subtracting the inverses of the smallest and largest eigenvalues according to

$$\underbrace{\beta + \sum_{i=1}^{n+1} \frac{1}{\bar{\lambda}_i}}_{= 0} + \underbrace{\frac{1}{\bar{\lambda}_{n+2}} - \frac{1}{\bar{\lambda}_1}}_{>0} = \beta + \sum_{i=2}^{n+2} \frac{1}{\bar{\lambda}_i} > 0,$$

the condition (4.67) is proven to satisfy (4.64) and, hence, a set of real closed-loop eigenvalues that satisfies (4.65)–(4.67) also satisfies (4.62)–(4.64).

In order to derive a procedure, the condition (4.67) can be unfold to obtain

$$\underbrace{\beta}_{>0} + \underbrace{\frac{1}{\bar{\lambda}_1} + \frac{1}{\bar{\lambda}_2} + \ldots + \frac{1}{\bar{\lambda}_{n+1}}}_{<0} = 0, \tag{4.68}$$

where $\bar{\lambda}_1$ should be the dominant eigenvalue. An upper bound for $\bar{\lambda}_1$ is given by

$$\bar{\lambda}_1 < -\frac{1}{\beta}, \tag{4.69}$$

which guarantees that the subtotal $\beta + \bar{\lambda}_1^{-1}$ is positive. Otherwise, (4.68) has no solution since all remaining eigenvalues should also be negative. Thus, for a fixed $\bar{\lambda}_1$, the requirement

$$\underbrace{\beta + \frac{1}{\bar{\lambda}_1}}_{>0} + \underbrace{\frac{1}{\bar{\lambda}_2} + \ldots + \frac{1}{\bar{\lambda}_{n+1}}}_{<0} = 0$$

has to be satisfied by choosing $\bar{\lambda}_2$ according to

$$\bar{\lambda}_2 < -\frac{1}{\beta + \bar{\lambda}_1^{-1}}. \tag{4.70}$$

This procedure can be repeated for the remaining eigenvalues which leads to the recursion rule

$$\bar{\lambda}_i < -\frac{1}{\beta + \sum_{j=1}^{i-1} \bar{\lambda}_j^{-1}}, \quad i = 2, 3, \ldots, n. \tag{4.71}$$

The last eigenvalue is determined by

$$\bar{\lambda}_{n+1} = -\frac{1}{\beta + \sum_{j=1}^{n} \bar{\lambda}_j^{-1}} \tag{4.72}$$

to meet the condition (4.67) exactly.

The procedure is illustrated in Fig. 4.9. The upper bounds are determined iteratively starting with the largest upper bound (4.69) for the dominant eigenvalue. The subsequent upper bounds (4.71) extend further and further into the left complex plane.

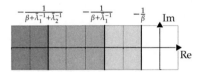

Fig. 4.9: Upper bounds of the closed-loop eigenvalues

Choosing the closed-loop eigenvalues according to this procedure will make sure that the requirement (4.67) is satisfied so that the objective (4.66) is achieved. In order to guarantee that $\bar{\lambda}_1$ is the dominant eigenvalue (to satisfy (4.65)), a lower bound has to be imposed: The upper bound (4.70) of $\bar{\lambda}_2$ has to be located to the left of $\bar{\lambda}_1$ in the complex plane so that there is no eigenvalue with a larger real part than $\bar{\lambda}_1$, i.e.

$$\bar{\lambda}_2 < -\frac{1}{\beta + \bar{\lambda}_1^{-1}} < \bar{\lambda}_1,$$

which is satisfied if and only if $-2\beta^{-1} < \bar{\lambda}_1$. Hence, the feasible region for $\bar{\lambda}_1$ to be the dominant eigenvalue is

$$-\frac{2}{\beta} < \bar{\lambda}_1 < -\frac{1}{\beta}. \tag{4.73}$$

The developed procedure to find closed-loop eigenvalues that render the controlled vehicle asymptotically stable and externally positive is summarised in Algorithm 4.1.

Remark. The eigenvalues obtained by Algorithm 4.1 only represent one possible solution to satisfy the requirements of Theorem 4.4. Another possibility is to place all closed-loop eigenvalues in the same location. The condition (4.64) is then satisfied if

$$\bar{\lambda}_i = -\frac{n+1}{\beta}, \quad i = 1, 2, \ldots, n+1$$

holds true. However, for a large system order n, the eigenvalues are shifted far to the left in the complex plane which restricts the range of feasible values for β if control input boundaries are considered. The eigenvalues determined by Algorithm 4.1 are more widely distributed on the real axis but the dominant eigenvalue $\bar{\lambda}_1$ is allowed to be slower, i.e. located closer to the origin of the complex plane which allows for implementing a smaller time-headway coefficient β.

Algorithm 4.1 (Eigenvalues that achieve external positivity).

Input: Desired time headway β and dynamical order n of Σ_i.

1 Choose the dominant eigenvalue $\bar{\lambda}_1$ from the interval (4.73).
2 **for** $i = 2, 3, \ldots, n$
3 $\quad$| $\quad$ Choose $\bar{\lambda}_i$ iteratively with (4.71).
4 **end**
5 Determine $\bar{\lambda}_{n+1}$ with (4.72).
6 Choose the last eigenvalue according to $\bar{\lambda}_{n+2} < \bar{\lambda}_{n+1}$.

Output: Eigenvalues $\bar{\lambda}_i, (i = 1, 2, \ldots, n+2)$ that satisfy (4.62)–(4.64).

4.3.6 Summary of the design procedure

Given the closed-loop eigenvalues found by Algorithm 4.1, the coefficients $\bar{a}_i$ of the characteristic polynomial are determined by (4.56). The feedback gain k that solves the ACC design problem is then determined by Ackermann's formula (4.39) as stated by the following theorem.

Theorem 4.5 (Solution to the ACC design problem). *If the eigenvalues $\bar{\lambda}_i, (i = 1, 2, \ldots, n+2)$ that are chosen according to Algorithm 4.1 are placed by the state feedback gain k, the controlled vehicle (4.34) satisfies the design objectives (D1)–(D4).*

Proof. Choosing the dominant eigenvalue from the interval (4.73) ensures that $\bar{\lambda}_1$ is real and negative and the other eigenvalues are located to the left of it due to the recursion rule (4.71). Hence, (4.62) and (4.63) are satisfied. Furthermore, the choice of the remaining eigenvalues $\bar{\lambda}_i, (i = 2, 3, \ldots, n+2)$ by (4.71), (4.72) and $\bar{\lambda}_{n+2} < \bar{\lambda}_{n+1}$ guarantees that condition (4.67) is satisfied which in turn satisfies (4.64). Thus, the controlled vehicle $\bar{\Sigma}_i$ has a positive decomposition and is asymptotically stable which proves the theorem. $\blacksquare$

The complete design procedure is summarised by Algorithm 4.2. The vehicle model (4.25) does not have to be identical for each vehicle i. A set of controlled vehicles combined in any order with controllers that are designed according to this procedure will satisfy the requirements (R1)–(R4).

Algorithm 4.2 (Design of externally positive vehicles).

Input: $\Sigma_i = (A, b, c^\mathsf{T})$ and desired time headway β.

1 Execute Algorithm 4.1 with $n = \dim A$ and β.
2 Calculate the coefficients $\bar{a}_i$, $(i = 0, 1, \ldots, n+1)$ with (4.56).
3 Construct A_0 and S_0 according to (4.32) and (4.42).
4 Determine k with Ackermann's formula (4.39).

Output: Feedback gain k with which $\bar{\Sigma}_i$ satisfies (D1)–(D4).

4.3.7 Application example

Consider a vehicle (4.25) with the transfer function

$$G(s) = \frac{1}{s\,(\tau s + 1)}$$

with the minimal state-space realisation

$$\Sigma_i : \begin{cases} \begin{pmatrix} \dot{v}_i(t) \\ \dot{a}_i(t) \end{pmatrix} = \begin{pmatrix} 0 & 1 \\ 0 & -\tau^{-1} \end{pmatrix} \begin{pmatrix} v_i(t) \\ a_i(t) \end{pmatrix} + \begin{pmatrix} 0 \\ \tau^{-1} \end{pmatrix} u_i(t) \\ v_i(t) = \begin{pmatrix} 1 & 0 \end{pmatrix} \begin{pmatrix} v_i(t) \\ a_i(t) \end{pmatrix}, \end{cases}$$

where $a_i(t)$ denotes the acceleration of the vehicle. This model satisfies Assumption 4.2 and is widely used in the literature of vehicle platooning (cf. [64, 99, 103, 114, 146] for example). The time constant for the test vehicles in [114] was identified as $\tau = 100\,\text{ms}$. The open-loop dynamics (4.32) read as

$$\begin{pmatrix} \dot{v}_i(t) \\ \dot{a}_i(t) \\ \dot{d}_i(t) \\ \dot{z}_i(t) \end{pmatrix} = \begin{pmatrix} 0 & 1 & 0 & 0 \\ 0 & -\tau^{-1} & 0 & 0 \\ -1 & 0 & 0 & 0 \\ \beta & 0 & -1 & 0 \end{pmatrix} \begin{pmatrix} v_i(t) \\ a_i(t) \\ d_i(t) \\ z_i(t) \end{pmatrix} + \begin{pmatrix} 0 \\ \tau^{-1} \\ 0 \\ 0 \end{pmatrix} u_i(t) + \begin{pmatrix} 0 \\ 0 \\ 1 \\ 0 \end{pmatrix} v_{i-1}(t). \qquad (4.74)$$

The procedure presented in this chapter will be applied to find a feedback gain k that renders the controlled vehicle asymptotically stable and externally positive via

$$u_i(t) = -k^\mathsf{T} \begin{pmatrix} v_i(t) \\ a_i(t) \\ d_i(t) \\ z_i(t) \end{pmatrix} \qquad (4.75)$$

for the time-headway coefficient $\beta = 1$ s. Note that the state consisting of the velocity $v_i(t)$, the acceleration $a_i(t)$, the inter-vehicle distance $d_i(t)$ as well as the regulator state $z_i(t)$ are quantities that are measurable in modern vehicles and, hence, there is no need for an observer or estimator.

Ackermann's formula (4.39) yields

$$k^{\mathrm{T}} = \begin{pmatrix} \bar{a}_0 & \bar{a}_1 & \bar{a}_2 & \bar{a}_3 & 1 \end{pmatrix} \begin{pmatrix} 0 & 0 & \beta\tau & \tau \\ 0 & 0 & -\tau & 0 \\ \tau & 0 & 0 & 0 \\ 0 & \tau & 0 & 0 \\ 0 & -1 & 0 & 0 \end{pmatrix}. \tag{4.76}$$

In order to determine the coefficients $\bar{a}_i$, Algorithm 4.1 is executed with $n = 2$ and $\beta = 1$ s as follows: At first, the dominant closed-loop eigenvalue $\bar{\lambda}_1$ is chosen from the interval $(-2, -1)$ as

$$\bar{\lambda}_1 = -1.5\,\mathrm{s}^{-1}.$$

This choice results in the boundary $\bar{\lambda}_2 < -3\,\mathrm{s}^{-1}$ which is satisfied by

$$\bar{\lambda}_2 = -5\,\mathrm{s}^{-1}.$$

The second to last eigenvalue is fixed by (4.72) as

$$\bar{\lambda}_3 = -7.5\,\mathrm{s}^{-1}$$

and the last eigenvalue, which is equal to the transmission zero, is chosen as

$$\bar{\lambda}_4 = \bar{\mu} = -9\,\mathrm{s}^{-1}.$$

Using (4.56), the coefficients $\bar{a}_i$ are calculated which fully determines the right-hand side of (4.76) and results in

$$k^{\mathrm{T}} = \begin{pmatrix} 18.225 & 1.300 & -5.625 & 50.625 \end{pmatrix}.$$

The impulse response $\bar{g}(t) \circ\!\!-\!\!\bullet \bar{G}(s)$ is shown in Fig. 4.10 (a) (solid) to verify that the controlled vehicle (4.34) is indeed externally positive.

Comparison with the ideal vehicle dynamics

Figure 4.10 shows a comparison of the ideal vehicle dynamics (4.8) introduced in the beginning of this chapter with the application example of this section. The nonnegative impulse responses are shown in Fig. 4.10 (a). The ideal impulse response starts from a nonzero value. This behaviour is not possible to achieve with real vehicles since it requires the acceleration to be discontinuous.

The velocity response in Fig. 4.10 (b) demonstrates the lag of the nonideal vehicle. In the beginning, the velocity of the ideal vehicle is larger and after 1 s, it is smaller than the velocity of the nonideal vehicle. This observation is due to the fact that both vehicles have the same delay $G_d(0) = \beta$ (cf. Definition 2.2) and, hence, the nonideal vehicle that is slower in the beginning has to be faster to the end to satisfy the objective (D3) on the delay.

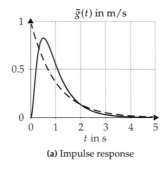

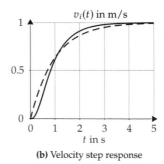

(a) Impulse response   **(b)** Velocity step response

Fig. 4.10: Comparison of the ideal vehicle dynamics (4.8) (dashed) with (4.74)–(4.76) (solid)

4.4 Summary and literature notes

This chapter presented a solution to the design of adaptive cruise controllers using a state-feedback approach. To this end, the requirements (R1)–(R4) on the desired behaviour of the platoon are formalised and the local design objectives (D1)–(D4) are derived that should be satisfied by all controlled vehicles. It has been shown that an appropriately chosen set of closed-loop eigenvalues that are placed with pole assignment techniques satisfies the design objectives with the proposed control structure.

Some results of this chapter have been published in conference and journal contributions by the author. The extended state-feedback structure in Fig. 4.8 has been proposed in [2, 4]. The fundamental requirement that the controlled vehicle has to be externally positive has been discussed in [3, 9] where an experimental comparison of $\mathcal{L}_2$ string stable and externally positive vehicles is given. The design of adaptive cruise controllers as the main result of this chapter has been published in [5] for a special vehicle model and in [8] for a general model.

Delaunay triangulation networks

5

This chapter addresses the maintenance of a communication network in which mobile agents are connected to their nearest neighbours. The Delaunay triangulation will be introduced as a proximity network and known properties are summarised. The main contributions of the author are given in Section 5.2 and Section 5.3 where an efficient method to test the validity of the network locally and distributed algorithms are proposed, respectively, that enable the agents to monitor and to switch communication links in real time so that the overall network structure is a Delaunay triangulation at any time.

5.1 The Delaunay triangulation

5.1.1 Introduction to proximity networks

Mobile agents as encountered in autonomous driving, future logistics or robot formation control have to cooperate with each other to achieve an overarching goal. Thus, the agents are equipped with communication systems that allow for exchanging locally available information. The agents are changing their positions represented by the Cartesian coordinates

$$p_i(t) = \begin{pmatrix} x_i(t) \\ y_i(t) \end{pmatrix}, \quad i = 1, 2, \ldots, N$$

according to their dynamics P_i as a result of individual control inputs generated by the local feedbacks F_i in Fig. 5.1. Due to the individual movement of the agents, a fixed communication structure eventually becomes unsuitable for the control task because the relative positions of the agents change over time. To overcome this problem, the communication structure should be adjusted by the communication unit D_i based on the current geometrical configuration of the agents while the agents are moving on arbitrary trajectories. Such a structure that considers the geometric configuration is called a *proximity network* [95].

The switching communication structure is represented by an undirected graph $\mathcal{G}(t) = (\mathcal{V}, \mathcal{E}(t))$ with a fixed set of vertices $\mathcal{V} = \{1, 2, \ldots, N\}$ representing the agents and a set

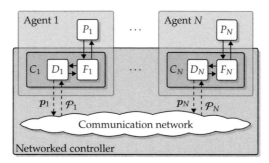

Fig. 5.1: Networked control structure of a multi-agent system

of edges $\mathcal{E}(t)$ representing the communication links. If there exists an edge $(i, j) \in \mathcal{E}(t)$, agents i and j are called *neighbours*. The set of all neighbours of an agent is defined as

$$\mathcal{N}_i(t) := \big\{ j \mid (i, j) \in \mathcal{E}(t) \big\}, \quad i \in \mathcal{V}.$$

Since the agents move relatively to one another, the set of neighbours $\mathcal{N}_i(t)$ is adapted throughout the movement of the agents and, thus, $\mathcal{N}_i(t)$ as well as $\mathcal{G}(t)$ and $\mathcal{E}(t)$ are switched at discrete points in time. The methods to be elaborated should enable each agent i to determine its set $\mathcal{N}_i(t)$ of neighbours so as to satisfy local conditions imposed on the communication network.

The *Delaunay triangulation* as a special proximity network will be proposed as the basis for detecting any change in the set of neighbours of an agent. This triangulation has two important properties: First, it yields a dense network that provides all necessary information to maintain itself with significantly less communication links compared to the complete graph [65]. A dense network that takes the proximity of the agents into account is useful for flocking [106] and allows for a quick propagation of information and, hence, a quick convergence of the dynamical behaviour of the agents [107]. Second, it can be maintained by local rules that can be applied by the individual agents separately without a global coordinator as will be elaborated in this chapter as the main contribution.

5.1.2 Definition of the Delaunay triangulation

Before the Delaunay triangulation is defined formally, some basic notions of computational geometry are introduced in this section. For the sake of comprehensibility, suppose that the positions $p_i, (i \in \mathcal{V})$ of the vertices are fixed preliminarily. Consider the set $\mathcal{P} := \{p_1, p_2, \ldots, p_N\}$ consisting of the positions of all agents and let $\mathcal{H}(\mathcal{P})$ denote the convex hull of $\mathcal{P}$. Furthermore, consider the following notions. Three or more points are called *collinear* if there exists a straight line that passes through the points. Furthermore, four or more points are called *cocircular* if there exists a circle that passes through all four points. The unique circle that passes through three noncollinear points is called the *circumcircle* and its centre is called the *circumcenter*. A triangulation is defined as follows.

Definition 5.1 (Triangulation [116]). A *triangulation* $\mathcal{T}(\mathcal{P})$ of a set $\mathcal{P}$ of points is a planar graph with vertices at the coordinates $p_i \in \mathcal{P}$ and edges $(i, j) \in \mathcal{E}$ that subdivide the convex hull $\mathcal{H}(\mathcal{P})$ into triangles so that the union of all triangles equals the convex hull.

Any triangulation $\mathcal{T}(\mathcal{P})$ with N vertices has

$$|\mathcal{E}| = 3\,(N-1) - M \tag{5.1}$$

edges where M denotes the number of agents on the boundary $\partial\mathcal{H}(\mathcal{P})$ of the convex hull [45, 72]. The edges of a triangulation do not cross each other and each face in the interior of $\mathcal{H}(\mathcal{P})$ is a triangle. There exists a triangulation for a set $\mathcal{P}$ if at least one point $p_i \in \mathcal{P}$ is noncollinear to the others. Furthermore, the triangulation of $N > 3$ points is not unique generally.

The Delaunay triangulation is a special triangulation which is named after its inventor Boris N. Delone (Delaunay is the French transliteration) who addressed this topic in the original paper [35]. It is a proximity graph that can be constructed by the geometrical configuration of the vertices. It is characterised by the following local property.

Definition 5.2 (Local Delaunay property [33, 35]). Consider a triangulation $\mathcal{T}(\mathcal{P})$ of a set $\mathcal{P}$ of points and let (A, B, C) denote a triangle with vertices at the coordinates p_i, $(i \in \{A, B, C\} \subseteq \mathcal{P})$. A triangle is said to be *Delaunay* if there is no point $p_i \in \mathcal{P}$ in the interior of the circumcircle through p_A, p_B and p_C.

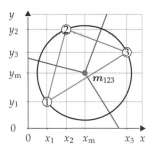

Fig. 5.2: A triangle and its circumcircle

As an example, consider the circumcircle of the triangle shown in Fig. 5.2. It can be constructed by extending the bisectors of the edges, which meet in the circumcenter $m_{123} = \begin{pmatrix} x_m & y_m \end{pmatrix}^{\mathsf{T}}$. According to Definition 5.2, the triangle spanned by p_1, p_2 and p_3 is Delaunay since there does not exist any point $p_i \in \mathcal{P}, (i \in \mathcal{V} \setminus \{1, 2, 3\})$ that is closer to the circumcenter than the vertices of the triangle (1,2,3).

Given the previous considerations, the Delaunay triangulation can be finally defined as follows.

Definition 5.3 (Delaunay triangulation [33, 35]). A *Delaunay triangulation* $\mathcal{DT}(\mathcal{P})$ is a triangulation whose triangles are all Delaunay.

5.1.3 Important properties of the Delaunay triangulation

There are several distinctive properties of the Delaunay triangulation. This section will focus on the properties that are helpful for the determination of a communication structure for multi-agent systems. Further properties of the Delaunay triangulation are given in the monographs [33, 52, 95, 116]. The most important property is an alternative characterisation of a Delaunay triangulation given by the following lemma.

Lemma 5.1 (Empty circle property [33]). *Two vertices $i, j \in \mathcal{V}$ at the positions $p_i, p_j \in \mathcal{P}$ of a Delaunay triangulation $\mathcal{DT}(\mathcal{P})$ are connected by an edge (i, j) if and only if there exists a circle passing through p_i and p_j that does not contain a point $p_k \in \mathcal{P}, (k \in \mathcal{V})$ in its interior.*

This property is illustrated in Fig. 5.3. The solid circle passes through the neighbouring vertices $i = 3$ and $j = 4$ and does not contain a point in its interior. Thus, the encircled edge (3, 4) is part of the Delaunay triangulation. This property allows to conclude that each agent is guaranteed to be connected to its geometrically closest neighbour in a Delaunay triangulation as illustrated by the dashed circle in Fig. 5.3 which is summarised by the following lemma.

Lemma 5.2 (Nearest neighbours [95]). *If the communication graph $\mathcal{G}$ is a Delaunay triangulation $\mathcal{DT}(\mathcal{P})$, each agent $i \in \mathcal{V}$ is connected to its geometrically closest neighbour:*

$$j^\star = \arg \min_{\substack{j \in \mathcal{V} \\ j \neq i}} \left\| p_i - p_j \right\| \implies j^\star \in \mathcal{N}_i.$$

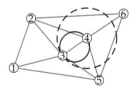

Fig. 5.3: Visualisation of Lemma 5.1 and Lemma 5.2

Lemma 5.2 provides a notable property considering mobile agents: In order to achieve collision avoidance among all agents, it is sufficient to achieve collision avoidance with neighbours if the Delaunay triangulation is maintained along the trajectories of a set of agents. This property makes the Delaunay triangulation an appropriate choice for a communication structure of a multi-agent system with mobile agents. To this end, this chapter contributes with the following results.

1. An efficient way to characterise the Delaunay triangulation geometrically from the perspective of an agent will be elaborated in Section 5.2.

2. Distributed algorithms to be performed locally on the agents by the decision units D_i (Fig. 5.1) will be developed in Section 5.3 that allow for maintaining the Delaunay triangulation while the agents are moving.

This chapter serves as the methodical fundament for Chapter 6 where the Delaunay triangulation will be applied to a vehicle formation problem.

5.2 Local geometric characterisation of the Delaunay triangulation

5.2.1 Metrics to test geometric configurations

In order to maintain the Delaunay triangulation in real time, an efficient way to test the current communication structure is presented in the following that allows for validating whether the local Delaunay property according to Definition 5.2 is satisfied. The goal is to derive local algorithms to be performed by the agents with which the agents can establish or remove edges from the network so that the network is a Delaunay triangulation at any time. To this end, this section will introduce different relations that are useful to characterise the Delaunay triangulation.

Orientation matrix

Consider three points p_i, $(i \in \{A, B, C\})$. The *orientation matrix* is defined as

$$O_{ABC} := \begin{pmatrix} x_A & y_A & 1 \\ x_B & y_B & 1 \\ x_C & y_C & 1 \end{pmatrix} \tag{5.2}$$

and allows for determining the relative position of the three points by evaluation of the sign of its determinant [45, 116]. If all three points are collinear, they have a common representation by the line equation

$$y_i = m\, x_i + y_0, \quad i \in \{A, B, C\} \tag{5.3}$$

with m and y_0 denoting the slope and the ordinate intercept of the line, respectively, which renders the columns of O_{ABC} linearly dependant. Thus, $|O_{ABC}| = 0$ if and only if the three considered points are collinear. Otherwise, the determinant is either negative or positive if the points are right bended or left bended, respectively, as in Fig. 5.4.

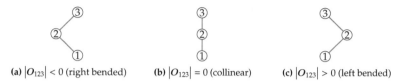

(a) $|O_{123}| < 0$ (right bended) (b) $|O_{123}| = 0$ (collinear) (c) $|O_{123}| > 0$ (left bended)

Fig. 5.4: Determination of the relative orientation

Note that $|O_{ABC}| = -|O_{CBA}|$ which means that the way the points are bended depends on the perspective and the ordering. Moreover, the determinant of the orientation matrix is invariant to circular shifting: $|O_{ABC}| = |O_{BCA}| = |O_{CAB}|$.

Circumcenter of a triangle

Three noncollinear points p_i, $(i \in \{A, B, C\})$ span a triangle whose *circumcenter* m_{ABC} is equally distanced to the points:

$$\left\| p_A - m_{ABC} \right\| = \left\| p_B - m_{ABC} \right\| = \left\| p_C - m_{ABC} \right\|. \tag{5.4}$$

The relation (5.4) can be cast in the system of equations

$$\left\| p_A \right\|^2 - \left\| p_B \right\|^2 - 2 \left(p_A - p_B \right)^\mathsf{T} m_{ABC} = 0 \tag{5.5}$$

$$\left\| p_B \right\|^2 - \left\| p_C \right\|^2 - 2 \left(p_B - p_C \right)^\mathsf{T} m_{ABC} = 0 \tag{5.6}$$

$$\left\| p_C \right\|^2 - \left\| p_A \right\|^2 - 2 \left(p_C - p_A \right)^\mathsf{T} m_{ABC} = 0 \tag{5.7}$$

by considering the pairwise differences. By introduction of the permutation matrix

$$P = \begin{pmatrix} 0 & 1 & 0 \\ 0 & 0 & 1 \\ 1 & 0 & 0 \end{pmatrix},$$

equations (5.5)–(5.7) can be reformulated equivalently as

$$(I - P) \left(\begin{pmatrix} \left\| p_A \right\|^2 \\ \left\| p_B \right\|^2 \\ \left\| p_C \right\|^2 \end{pmatrix} - 2 \begin{pmatrix} p_A^\mathsf{T} \\ p_B^\mathsf{T} \\ p_C^\mathsf{T} \end{pmatrix} m_{ABC} \right) = 0 \tag{5.8}$$

with $\mathbf{0}$ denoting the vector consisting of zeros only. The following lemma establishes a connection between the circumcenter m_{ABC} and the orientation matrix O_{ABC}.

Lemma 5.3 (Circumcenter). *Consider three noncollinear points p_A, p_B and p_C. Their circumcenter m_{ABC} is determined by*

$$\begin{pmatrix} m_{ABC} \\ \gamma \end{pmatrix} = \frac{1}{2} O_{ABC}^{-1} \begin{pmatrix} \|p_A\|^2 \\ \|p_B\|^2 \\ \|p_C\|^2 \end{pmatrix} \tag{5.9}$$

with O_{ABC} defined in (5.2) and γ denoting a scalar extension.

Proof. The matrix $I - P$ is a Laplacian matrix since it is singular and all row sums are equal to zero. Thus, its right eigenvector corresponding to the eigenvalue equal to zero is the vector consisting only of ones denoted by $\mathbf{1}$, i.e.

$$2\gamma (I - P) \mathbf{1} = 0 \tag{5.10}$$

holds true for any γ. Comparison of (5.8) with (5.10) results in

$$\begin{pmatrix} \|p_A\|^2 \\ \|p_B\|^2 \\ \|p_C\|^2 \end{pmatrix} - 2 \begin{pmatrix} p_A^{\mathsf{T}} \\ p_B^{\mathsf{T}} \\ p_C^{\mathsf{T}} \end{pmatrix} m_{ABC} = 2\gamma \, \mathbf{1},$$

which yields the relation

$$\begin{pmatrix} \|p_A\|^2 \\ \|p_B\|^2 \\ \|p_C\|^2 \end{pmatrix} = 2 \begin{pmatrix} p_A^{\mathsf{T}} & 1 \\ p_B^{\mathsf{T}} & 1 \\ p_C^{\mathsf{T}} & 1 \end{pmatrix} \begin{pmatrix} m_{ABC} \\ \gamma \end{pmatrix} = 2 \, O_{ABC} \begin{pmatrix} m_{ABC} \\ \gamma \end{pmatrix}$$

after rearrangement. The result (5.9) follows after reformulation. ∎

The result (5.9) allows for determining the circumcenter m_{ABC} straightforwardly for a given set of three noncollinear points. The scalar γ is a by-product that results from the fact that the system of equations (5.5)–(5.7) is redundant, i. e. two of those equations are sufficient to determine the circumcenter m_{ABC}. However, the compact representation (5.9) will be shown to be useful when testing the local Delaunay property.

As a preliminary work, consider a partition of the adjugate of O_{ABC} as

$$\mathrm{adj}\, O_{ABC} = \begin{pmatrix} \Lambda \\ \lambda^{\mathsf{T}} \end{pmatrix} \tag{5.11}$$

determined by the matrix

$$\Lambda = \begin{pmatrix} y_B - y_C & y_C - y_A & y_A - y_B \\ x_C - x_B & x_A - x_C & x_B - x_A \end{pmatrix}$$

representing the first two rows and the vector

$$\lambda^T = \left(\begin{vmatrix} x_B & y_B \\ x_C & y_C \end{vmatrix} \begin{vmatrix} x_C & y_C \\ x_A & y_A \end{vmatrix} \begin{vmatrix} x_A & y_A \\ x_B & y_B \end{vmatrix} \right)$$

representing the last row of the adjugate (5.11), respectively. The result (5.9) is then equivalently stated by the relations

$$2|O_{ABC}|\, m_{ABC} = \Lambda \begin{pmatrix} \|p_A\|^2 \\ \|p_B\|^2 \\ \|p_C\|^2 \end{pmatrix}, \qquad 2|O_{ABC}|\, \gamma = \lambda^T \begin{pmatrix} \|p_A\|^2 \\ \|p_B\|^2 \\ \|p_C\|^2 \end{pmatrix}, \tag{5.12}$$

which will be used in the next section to prove Lemma 5.4.

Enclosure matrix

Given four points p_i, ($i \in \{A, B, C, D\}$), the *enclosure matrix* is defined as

$$M_{ABCD} := \begin{pmatrix} x_A & y_A & x_A^2 + y_A^2 & 1 \\ x_B & y_B & x_B^2 + y_B^2 & 1 \\ x_C & y_C & x_C^2 + y_C^2 & 1 \\ x_D & y_D & x_D^2 + y_D^2 & 1 \end{pmatrix} \tag{5.13}$$

in [45, 52] as a further measure for the relative characterisation of different points. If all four points are cocircular, they have the common representation by the circle equation

$$x_i^2 + y_i^2 - 2x_m x_i - 2y_m y_i + x_m^2 + y_m^2 - r^2 = 0, \quad i \in \{A, B, C, D\} \tag{5.14}$$

with x_m, y_m and r denoting the coordinates of the middle point and the radius of the circle, respectively. Furthermore, if all four points are collinear, they have the common representation by the line equation (5.3). Both cases render the enclosure matrix M_{ABCD} singular and, hence, $|M_{ABCD}| = 0$ if and only if all four points are either cocircular or collinear.

5.2.2 Local Delaunay test

The matrices O_{ABC} and M_{ABCD} can be used to test whether a point is enclosed by the circumcircle of three other points as will be shown in this section. For that purpose, it will be assumed that the points p_A, p_B and p_C are fixed and it should be tested whether a variable candidate point p_D is encircled by the circumcircle through p_A, p_B and p_C. To avoid trivial solutions, it is assumed that p_A, p_B and p_C are noncollinear. The approach is as follows: It will be shown that $|M_{ABCD}|$ switches its sign if p_D crosses the circumcircle through p_A, p_B and p_C. Furthermore, it will be shown that the way the sign of $|M_{ABCD}|$ changes depends upon $|O_{ABC}|$. To this end, the following lemma establishes a connection between the determinants of O_{ABC} and M_{ABCD}.

Lemma 5.4 (Enclosure paraboloid). *For three fixed noncollinear points p_A, p_B, p_C and a variable candidate point p_D, the determinant of the enclosure matrix (5.13) is represented by the quadratic form*

$$|M_{ABCD}| = -|O_{ABC}| \left(p_D^T p_D - 2p_D^T m_{ABC} - 2\gamma \right), \qquad (5.15)$$

which describes a paraboloid with p_D as the argument.

Proof. Application of the Laplace expansion of the last row of M_{ABCD} yields

$$|M_{ABCD}| = -(x_D^2 + y_D^2) \begin{vmatrix} x_A & y_A & 1 \\ x_B & y_B & 1 \\ x_C & y_C & 1 \end{vmatrix} - x_D \begin{vmatrix} y_A & x_A^2 + y_A^2 & 1 \\ y_B & x_B^2 + y_B^2 & 1 \\ y_C & x_C^2 + y_C^2 & 1 \end{vmatrix}$$

$$+ y_D \begin{vmatrix} x_A & x_A^2 + y_A^2 & 1 \\ x_B & x_B^2 + y_B^2 & 1 \\ x_C & x_C^2 + y_C^2 & 1 \end{vmatrix} + \begin{vmatrix} x_A & y_A & x_A^2 + y_A^2 \\ x_B & y_B & x_B^2 + y_B^2 \\ x_C & y_C & x_C^2 + y_C^2 \end{vmatrix}.$$

The first minor is equal to $|O_{ABC}|$ and the remaining three minors are expanded along the columns with the quadratic terms which results in

$$|M_{ABCD}| = -p_D^T p_D |O_{ABC}| + p_D^T \underbrace{\begin{pmatrix} y_B - y_C & y_C - y_A & y_A - y_B \\ x_C - x_B & x_A - x_C & x_B - x_A \end{pmatrix}}_{\Lambda} \begin{pmatrix} \|p_A\|^2 \\ \|p_B\|^2 \\ \|p_C\|^2 \end{pmatrix}$$

$$+ \underbrace{\left(\begin{vmatrix} x_B & y_B \\ x_C & y_C \end{vmatrix} \quad \begin{vmatrix} x_C & y_C \\ x_A & y_A \end{vmatrix} \quad \begin{vmatrix} x_A & y_A \\ x_B & y_B \end{vmatrix} \right)}_{\lambda^T} \begin{pmatrix} \|p_A\|^2 \\ \|p_B\|^2 \\ \|p_C\|^2 \end{pmatrix}.$$

The result (5.15) follows by using the identities (5.12).   ∎

If p_D is located on the circumcircle through p_A, p_B and p_C, the determinant of M_{ABCD} is equal to zero as already shown by the circle equation (5.14) and the paragraph afterwards. Thus, the circumcircle is the level line of the paraboloid for $|M_{ABCD}| = 0$. Furthermore, its extreme point is located at the circumcenter m_{ABC} which can be concluded from the gradient of the determinant of M_{ABCD} given by

$$\nabla |M_{ABCD}| = -2 |O_{ABC}| (p_D - m_{ABC})$$

and illustrated in Fig. 5.5 for a variable p_D. The gradient is locally orthogonal to the circumcircle and, hence, the sign of the determinant of M_{ABCD} switches if p_D crosses the circumcircle. These observations lead to the first main result of this chapter which presents a method to test the local Delaunay property.

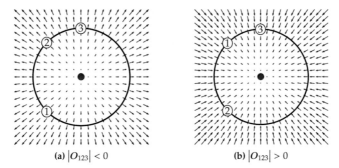

(a) $|O_{123}| < 0$ (b) $|O_{123}| > 0$

Fig. 5.5: The gradient $\nabla |M_{123D}|$ is radially directed away from or towards the circumcenter

> **Theorem 5.1** (Local Delaunay test). *Consider three fixed noncollinear points* p_A, p_B, p_C *and a variable candidate point* p_D. *The point* p_D *is enclosed by the circumcircle through* p_A, p_B *and* p_C *if and only if*
>
> $$\text{sgn} \left| M_{ABCD} \right| = \text{sgn} \left| O_{ABC} \right|.$$
>
> *Otherwise, the four points are either cocircular if* M_{ABCD} *is singular or the point* p_D *is outside of the circumcircle.*

Proof. The Hessian matrix of $\left| M_{ABCD} \right|$ is given by

$$H \left(\left| M_{ABCD} \right| \right) = -2 \left| O_{ABC} \right| I.$$

Thus, $\left| M_{ABCD} \right|$ is either a convex or a concave paraboloid with respect to a candidate point p_D if the determinant of O_{ABC} is negative or positive, respectively. Consequently, $\left| M_{ABCD} \right|$ has the same sign as $\left| O_{ABC} \right|$ if and only if p_D is in the interior of the circumcircle through p_A, p_B and p_C. ∎

Theorem 5.1 provides an efficient method to test the local Delaunay property using the points $p_i, (i \in \{A, B, C, D\})$ by comparing the signs of the determinants of the orientation matrix O_{ABC} and the enclosure matrix M_{ABCD}. This method can be executed by the agents that are located at the positions $p_i, (i \in \mathcal{V})$ using the positions of their respective neighbours $j \in \mathcal{N}_i$ to test whether the network is a Delaunay triangulation locally. If the test shows that the network is *not* a Delaunay triangulation, the agents have to take actions to change the network in real time so that the network is a Delaunay triangulation again. The following section will elaborate algorithms that perform this task of *maintaining* the Delaunay triangulation.

5.3 Distributed algorithms to maintain a Delaunay triangulation

5.3.1 The Lawson flip

The application of Theorem 5.1 to test whether a triangle is Delaunay is illustrated in Fig. 5.6. For that, consider the *mobile* agent $D = 4$. From left to right, the agent $D = 4$ enters the circumcircle through the agents $A = 1$, $B = 2$ and $C = 3$. Each frame is uniquely characterised by the signs of $\left| O_{ABC} \right|$ and $\left| M_{ABCD} \right|$. At first, the graph is a Delaunay triangulation. The edge $(2, 3)$ is called *valid* since it is the common edge of two triangles that are Delaunay since there is no other agent in the interior of their circumcircles. In the second frame, agent $D = 4$ is on the circumcircle and, hence, the Delaunay triangulation is not unique but the local property is still satisfied. In the last frame, agent $D = 4$ entered the circumcircle with the consequence that both triangles are not Delaunay according to Theorem 5.1 which is referred to as a *topological event* [19, 41, 130, 148].

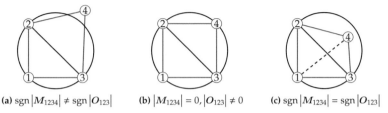

(a) sgn $|M_{1234}| \neq$ sgn $|O_{123}|$ (b) $|M_{1234}| = 0, |O_{123}| \neq 0$ (c) sgn $|M_{1234}| =$ sgn $|O_{123}|$

Fig. 5.6: Detection of a topological event using Theorem 5.1

After a topological event, the current network is not a Delaunay triangulation because the local Delaunay property is violated, i. e. at least one edge is not valid. To restore the local Delaunay property, the edge $(2, 3)$ has to be replaced by the edge $(1, 4)$ (dashed line in Fig. 5.6 (c)). This procedure of switching the common edge of two adjacent triangles that form a convex quadrilateral is known as the *Lawson flip* named after the researcher Charles L. Lawson who proved that a given triangulation can be transformed to any other triangulation by edge flipping [71]. Flipping according to the procedure illustrated in Fig. 5.6 will ensure that the local Delaunay property is satisfied after the flip so that the result is a Delaunay triangulation again [70]. Note that the Lawson flip can only be applied to edges which are the common edge of two adjacent triangles. Thus, the edges on the boundary of the convex hull have to be dealt with separately by other procedures that will be proposed in Section 5.3.4.

5.3.2 Local information

The rest of this chapter will consider mobile agents that are allowed to move relatively to one another, i. e. their positions $p_i(t)$, $(i \in \mathcal{V})$ evolve along continuous trajectories. The goal of the algorithms to be elaborated is to maintain the network so that it is a Delaunay triangulation. Thus, the set $\mathcal{N}_i(t)$ of neighbours as well as $\mathcal{G}(t)$ and $\mathcal{E}(t)$ are switched at discrete points in time. In order to maintain the proximity graph with the Lawson flip without a centralised unit while the agents are moving, each agent i stores local information in the set

$$\mathcal{I}_i(t) := \{p_i(t), \mathcal{N}_i(t), \mathcal{H}_i(t), \mathcal{T}_i(t)\},$$

whose elements are discussed in the following. In addition to the Cartesian position $p_i(t)$ and the set of neighbours $\mathcal{N}_i(t)$, the local information $\mathcal{I}_i(t)$ includes the set

$$\mathcal{H}_i(t) := \begin{cases} \{j \mid (i, j) \in \partial \mathcal{H}(\mathcal{P}(t))\}, & \text{if } p_i(t) \in \partial \mathcal{H}(\mathcal{P}(t)) \\ \emptyset, & \text{otherwise,} \end{cases}$$

which is either empty or contains the two adjacent neighbours on the boundary of the network denoted by $\partial \mathcal{H}(\mathcal{P}(t))$ if agent i is also on the boundary, and the set

$$\mathcal{T}_i(t) := \{\{j,k\} \mid (i,j), (i,k), (j,k) \in \mathcal{E}(t), \text{the triangle } (i,j,k) \text{ is Delaunay}\} \quad (5.16)$$

of the pairs of neighbours of all triangles of which agent i is a vertex. The additional condition in (5.16) that the triangle (i,j,k) has to be Delaunay is imposed to avoid that a triangle that is composed of multiple smaller triangles is added to the set $\mathcal{T}_i(t)$ as illustrated by the following example.

Example 5.1 (Local information). Consider the network depicted in Fig. 5.7 with $N = 7$ agents represented by the vertices. The locally available information $\mathcal{I}_i$ of each agent is given in Table 5.1.

Table 5.1: Local information $\mathcal{I}_i$ of the triangulation in Fig. 5.7

i	$\mathcal{N}_i$	$\mathcal{H}_i$	$\mathcal{T}_i$
1	$\{2,3,4,5\}$	$\emptyset$	$\{\{2,4\},\{3,4\},\{2,5\},\{3,5\}\}$
2	$\{1,4,5\}$	$\{5,4\}$	$\{\{1,4\},\{1,5\}\}$
3	$\{1,4,5,6,7\}$	$\{4,7\}$	$\{\{1,4\},\{1,5\},\{5,6\},\{6,7\}\}$
4	$\{1,2,3\}$	$\{2,3\}$	$\{\{1,2\},\{1,3\}\}$
5	$\{1,2,3,6,7\}$	$\{7,2\}$	$\{\{1,2\},\{1,3\},\{3,6\},\{6,7\}\}$
6	$\{3,5,7\}$	$\emptyset$	$\{\{3,5\},\{3,7\},\{5,7\}\}$
7	$\{3,5,6\}$	$\{3,5\}$	$\{\{3,6\},\{5,6\}\}$

Fig. 5.7: Example triangulation

With the local information $\mathcal{I}_i$, each agent knows whether it is part of the convex hull $\partial \mathcal{H}(\mathcal{P})$ and it can determine with which neighbours it forms a triangle. The triangle $(3,5,7)$ is *not* Delaunay and, hence, it is not contained in the sets $\mathcal{T}_i, (i \in \{3,5,7\})$. ☐

Initial information. It is assumed that the initial communication graph among the agents represents a Delaunay triangulation. One possibility to obtain an initial Delaunay triangulation is to use incremental algorithms as presented in [72] which, however, would require a coordinating unit temporarily. An algorithm to generate a Delaunay triangulation from scratch without a coordinator is not known yet and will not be included in this thesis. Thus, the initial information $\mathcal{I}_i(0)$ of each agent is assumed to be known and obtained by a possibly centralised algorithm.

Information obtained by communication. In addition to the locally stored informa-
tion $\mathcal{I}_i(t)$, the agents are able to receive the information $\mathcal{P}_i(t) \subset \mathcal{P}(t)$ with

$$\mathcal{P}_i(t) := \{ p_j(t) \mid j \in \mathcal{N}_i(t) \}$$

via digital communication over the network (cf. Fig. 5.1). Furthermore, the network is also
used to transmit and receive instructions as will be discussed in the following sections.

Note that the time symbol t of all quantities will be omitted for the sake readability in
the following sections.

5.3.3 Distributed implementation of the Lawson flip

The following investigation looks at the current network from the viewpoint of an agent $i \in$
$\mathcal{V}$. In regular time intervals, the agent checks all of its edges to its neighbours $j \in \mathcal{N}_i$.
For an edge (i, j), the agents C and D denote the other neighbours of the two adjacent
triangles as in Fig. 5.8 which can be identified using the local set of triangles $\mathcal{T}_i$ defined in
(5.16) as follows. Agent i searches among the elements of $\mathcal{T}_i$ for the two pairs

$$\{j, C\}, \{j, D\} \in \mathcal{T}_i \tag{5.17}$$

that both contain j. Two cases have to be considered: First, if there exist two elements that
both contain j, then the two corresponding agents C and D in (5.17) satisfy the structure
in Fig. 5.8. Second, if there is only one element in $\mathcal{T}_i$ that contains j, the edge (i, j) is on
the boundary of the network and, hence, does not have to be tested with Theorem 5.1.
Note that it is not possible that $\mathcal{T}_i$ has three elements that contain j since the Delaunay
triangulation is a planar graph.

Fig. 5.8: The condition $\mathrm{sgn} \left| M_{ijCD} \right| = \mathrm{sgn} \left| O_{ijC} \right|$ holds true

In order to check the validity of the edge (i, j), agent i has to evaluate the determinants
of M_{ijCD} and O_{ijC} according to Theorem 5.1. To this end, agent i has to gather information
about the Cartesian positions p_j, p_C and p_D via the communication network and compares
the signs of $\left| M_{ijCD} \right|$ and $\left| O_{ijC} \right|$. If they are different, p_D is not in the interior of the
circumcircle through p_i, p_j and p_C and, hence, the edge (i, j) is valid meaning that the
triangles (i, j, C) and (i, j, D) are both Delaunay. Otherwise, $\mathrm{sgn} \left| M_{ijCD} \right| = \mathrm{sgn} \left| O_{ijC} \right|$

holds true and p_D is in the circumcircle through p_i, p_j and p_C or, equivalently, p_C is in the circumcircle through p_i, p_j and p_D as illustrated in Fig. 5.8. Hence, the Lawson flip has to be applied to update the triangulation. For that purpose, the sets of local neighbours and triangles have to be updated according to Table 5.2. The first column contains locally performed updates and the remaining columns are sent as instructions to the respective neighbours via the communication network. The procedure is summarised in Algorithm 5.1. It has to be applied by all agents to each edge (i, j) that is *not* on the boundary $\partial \mathcal{H}(\mathcal{P})$ of the convex hull.

Table 5.2: Update of the individual information due to the Lawson flip

Agent i	Agent j	Agent C	Agent D
$N_i \leftarrow N_i \setminus \{j\}$	$N_j \leftarrow N_j \setminus \{i\}$	$N_C \leftarrow N_C \cup \{D\}$	$N_D \leftarrow N_D \cup \{C\}$
$\mathcal{T}_i \leftarrow \mathcal{T}_i \setminus \{j, C\}$	$\mathcal{T}_j \leftarrow \mathcal{T}_j \setminus \{i, C\}$	$\mathcal{T}_C \leftarrow \mathcal{T}_C \setminus \{i, j\}$	$\mathcal{T}_D \leftarrow \mathcal{T}_D \setminus \{i, j\}$
$\mathcal{T}_i \leftarrow \mathcal{T}_i \setminus \{j, D\}$	$\mathcal{T}_j \leftarrow \mathcal{T}_j \setminus \{i, D\}$	$\mathcal{T}_C \leftarrow \mathcal{T}_C \cup \{i, D\}$	$\mathcal{T}_D \leftarrow \mathcal{T}_D \cup \{i, C\}$
$\mathcal{T}_i \leftarrow \mathcal{T}_i \cup \{C, D\}$	$\mathcal{T}_j \leftarrow \mathcal{T}_j \cup \{C, D\}$	$\mathcal{T}_C \leftarrow \mathcal{T}_C \cup \{j, D\}$	$\mathcal{T}_D \leftarrow \mathcal{T}_D \cup \{j, C\}$

Algorithm 5.1 (Lawson flip).

Input: Local information $\mathcal{I}_i = \{p_i, N_i, \mathcal{H}_i, \mathcal{T}_i\}$ and $j \in N_i$.

1 Identify the two neighbours C and D which are the vertices of the triangles with the common edge (i, j) using $\mathcal{T}_i$.
2 Gather the current positions p_j, p_C and p_D via communication and construct M_{ijCD} and O_{ijC}.
3 if sgn $|M_{ijCD}|$ = sgn $|O_{ijC}|$
4 | Perform and send instructions according to Table 5.2.
5 end

Output: Updated local information $\mathcal{I}_i$.

Figure 5.9 shows an example of the execution of Algorithm 5.1. Initially, the network is a Delaunay triangulation. Then, agent $j = 6$ moves away from the other agents in Fig. 5.9 (a). Agent $i = 1$ checks the edge $(1, 6)$ by gathering the positions of the relevant agents, i.e. p_2, p_5 and p_6 are transmitted to agent $i = 1$ (depicted by dashed arrows) which checks the local Delaunay property according to Theorem 5.1 (Fig. 5.9 (b)). The test shows that the circumcircle through p_1, p_5 and p_6 encircles the point p_2 and, hence, the edge $(1, 6)$ has to be removed and the new edge $(2, 5)$ has to be introduced to the network. Consequently, agent $i = 1$ sends instructions to the involved agents in Fig. 5.9 (c). The labels $\cup\{2\}$ and $\setminus\{1\}$ are understood as "include agent $D = 2$" and "exclude agent $i = 1$", respectively, and represent the corresponding columns of Table 5.2. By executing the instructions, the edge

$(1, 6)$ is removed and the edge $(2, 5)$ is introduced to the network resulting in the Delaunay triangulation in Fig. 5.9 (d).

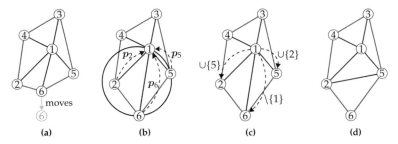

Fig. 5.9: Steps of the execution of Algorithm 5.1 by agent $i = 1$

5.3.4 Procedure for the convex hull

The edges on the boundary of the convex hull have to be tested separately for two cases:

1. An agent can move from the interior to the boundary of the convex hull. Due to the relation (5.1), an edge has to be removed from the network since the number of agents on the boundary has increased.

2. An agent can move from the boundary to the interior in which case an edge has to be added to the network.

(a) First case   (b) Second case

Fig. 5.10: Two possible cases at the boundary of the network

In Fig. 5.10, the edge (i, C) has to be removed (first case) or the edge (C, D) has to be added (second case). The agents C and D denote the next neighbours counterclockwise and clockwise, respectively, on the boundary. In order to adapt the Delaunay triangulation correctly, it will be utilised that an agent on the boundary of the network has exactly two edges (i, C) and (i, D) with $C, D \in \mathcal{H}_i$ that are also on the boundary. Hence, if $\mathcal{H}_i \neq \emptyset$ holds true, agent i has to test both adjacent edges on the boundary. The methods presented in this section use the Jarvis march (cf. [55, 116]) to test whether an agent is on the boundary of the convex hull.

First case

As illustrated in Fig. 5.10 (a), the first case describes the situation that an agent j in the interior of the network moves to the boundary. Without loss of generality, suppose that agent j moves towards the edge (i, C). As soon as it reaches the boundary, the edge (i, C) that it is surpassing has to be removed from the network which is executed by agent i in this example. For that, agent i tests the sign of $|O_{ijC}|$. If it is positive, agent j surpassed the edge (i, C). Agent i then executes and sends instructions according to Table 5.3 to remove the edge (i, C) from the network. The procedure is summarised in Algorithm 5.2 which is executed for each neighbour $j \in \mathcal{N}_i$ that is *not* an element of the set $\mathcal{H}_i$, i.e. the edge (i, j) is not on the boundary of the network.

Table 5.3: Instructions to update the convex hull (first case)

Agent i	Agent C	Agent j
$\mathcal{N}_i \leftarrow \mathcal{N}_i \setminus \{C\}$	$\mathcal{N}_C \leftarrow \mathcal{N}_C \setminus \{i\}$	
$\mathcal{T}_i \leftarrow \mathcal{T}_i \setminus \{j, C\}$	$\mathcal{T}_C \leftarrow \mathcal{T}_C \setminus \{i, j\}$	$\mathcal{T}_j \leftarrow \mathcal{T}_j \setminus \{i, C\}$
$\mathcal{H}_i \leftarrow \mathcal{H}_i \setminus \{C\}$	$\mathcal{H}_C \leftarrow \mathcal{H}_C \setminus \{i\}$	$\mathcal{H}_j \leftarrow \mathcal{H}_j \cup \{i, C\}$
$\mathcal{H}_i \leftarrow \mathcal{H}_i \cup \{j\}$	$\mathcal{H}_C \leftarrow \mathcal{H}_C \cup \{j\}$	

Algorithm 5.2 (Update of the convex hull (first case)).

Input: Local information $\mathcal{I}_i = \{p_i, \mathcal{N}_i, \mathcal{H}_i, \mathcal{T}_i\}$ and $j \in \mathcal{N}_i$.

1 Gather the positions p_j and p_C with $C \in \mathcal{H}_i$ via communication and construct the matrix O_{ijC}.
2 **if** $|O_{ijC}| > 0$
3   Perform and send instructions according to Table 5.3.
4 **end**

Output: Updated local information $\mathcal{I}_i$.

Figure 5.11 shows an example of the execution of Algorithm 5.2. Again, the network represents a Delaunay triangulation initially. Agent $j = 1$ moves from the interior to the boundary of the network as it surpasses the edge $(5, 3)$ in Fig. 5.11 (a). Agent $i = 5$ tests the edge $(5, 3)$ by evaluating the sign of $|O_{513}|$. For that, agent $i = 5$ receives the positions p_1 and p_3 via communication (Fig. 5.11 (b)). Since $|O_{513}| > 0$ (cf. Fig. 5.4) holds true, the edge $(5, 3)$ has to be removed from the network. Thus, agent $i = 5$ sends the corresponding instruction to agent $C = 3$ as in Fig. 5.11 (c) and after execution, the network is a Delaunay triangulation again (Fig. 5.11 (d)).

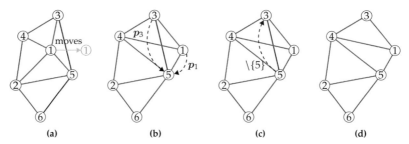

(a)   (b)   (c)   (d)

Fig. 5.11: Example of the execution of Algorithm 5.2 by Agent $i = 5$

Second case

The second case illustrated in Fig. 5.10 (b) describes the situation that an agent i on the boundary of the network moves to the interior. As soon as it leaves the boundary, the edge (C, D) has to be added to the network. For that, agent i tests the sign of $|O_{iCD}|$. If it is negative, agent i reached the interior of the network and the edge (C, D) has to be added in order to include the convex hull. Thus, agent i executes and sends instructions according to Table 5.4 which will add the edge (C, D) to the network. The procedure is summarised in Algorithm 5.3.

Table 5.4: Instructions to update the convex hull (second case)

Agent i	Agent C	Agent D
	$\mathcal{N}_C \leftarrow \mathcal{N}_C \cup \{D\}$	$\mathcal{N}_D \leftarrow \mathcal{N}_D \cup \{C\}$
$\mathcal{T}_i \leftarrow \mathcal{T}_i \cup \{C, D\}$	$\mathcal{T}_C \leftarrow \mathcal{T}_C \cup \{i, D\}$	$\mathcal{T}_D \leftarrow \mathcal{T}_D \cup \{i, C\}$
$\mathcal{H}_i \leftarrow \emptyset$	$\mathcal{H}_C \leftarrow \mathcal{H}_C \setminus \{i\}$	$\mathcal{H}_D \leftarrow \mathcal{H}_D \setminus \{i\}$
	$\mathcal{H}_C \leftarrow \mathcal{H}_C \cup \{D\}$	$\mathcal{H}_D \leftarrow \mathcal{H}_D \cup \{C\}$

Figure 5.12 shows an example of the execution of Algorithm 5.3. The initial network represents a Delaunay triangulation. Agent $i = 4$ moves from the boundary to the interior of the network in Fig. 5.12 (a). It tests whether it is still part of the boundary by evaluating the sign of $|O_{423}|$. For that, agent $i = 4$ receives the positions p_2 and p_3 via the current communication links (Fig. 5.12 (b)). Since $|O_{423}| < 0$ holds true (cf. Fig. 5.4), the edge $(2, 3)$ has to be added to the network. Thus, agent $i = 4$ sends the corresponding instructions to agents $C = 2$ and $D = 3$ in Fig. 5.12 (c) and after execution, the network is a Delaunay triangulation again (Fig. 5.12 (d)).

Algorithm 5.3 (Update of the convex hull (second case)).

Input: Local information $\mathcal{I}_i = \{p_i, \mathcal{N}_i, \mathcal{H}_i, \mathcal{T}_i\}$ and $j \in \mathcal{N}_i$.

1 Gather the positions p_C and p_D with $C, D \in \mathcal{H}_i$ via communication and construct the matrix O_{iCD}.
2 **if** $|O_{iCD}| < 0$
3    $\quad$ Perform and send instructions according to Table 5.4.
4 **end**

Output: Updated local information $\mathcal{I}_i$.

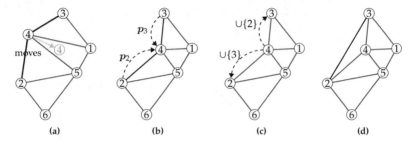

(a) (b) (c) (d)

Fig. 5.12: Example of the execution of Algorithm 5.3 by Agent $i = 4$

5.3.5 Summary of the procedure

The algorithms presented in the previous sections are summarised in Algorithm 5.4 which is executed by each agent i individually. For practical applications in a discrete-time realisation, it is assumed that the algorithm is invoked after each sampling period at the time instances

$$t = k\,T_{\text{s}}, \quad k = 0, 1, 2, \ldots$$

with T_{s} denoting the sampling time.

The Lawson flip procedure is only able to update the network correctly if it is a triangulation [71]. However, the two possible cases at the boundary $\partial\mathcal{H}(\mathcal{P})$ of the convex hull have as a consequence that some edges intersect each other (first case) or that the convex hull is not part of the network locally (second case) as illustrated in Fig. 5.10. Thus, if agent i is on the boundary of the convex hull ($\mathcal{H}_i \neq \emptyset$), Algorithms 5.2 and 5.3 are executed first to ensure that the communication graph is a triangulation before the Lawson flip is executed by invoking Algorithm 5.1. The correctness of Algorithm 5.4 can be deduced from the correctness of the Lawson flip algorithm and the Jarvis march which

---

Algorithm 5.4 (Local algorithm on agent $i = 1, 2, \ldots, N$).

Input: Initial local information $\mathcal{I}_i(0) = \{p_{i0}, \mathcal{N}_{i0}, \mathcal{H}_{i0}, \mathcal{T}_{i0}\}$.

1 **for** $k = 0, 1, 2, \ldots$
2   **if** $\mathcal{H}_i = \emptyset$
3   **for all** $j \in \mathcal{N}_i$
4           $\mathcal{I}_i \leftarrow$ Algorithm 5.1 (Lawson flip).
5   **end**
6   **else**
7        $\mathcal{I}_i \leftarrow$ Algorithm 5.3 (Convex hull, 2nd case).
8   **for all** $j \in \mathcal{N}_i, j \notin \mathcal{H}_i$
9           $\mathcal{I}_i \leftarrow$ Algorithm 5.2 (Convex hull, 1st case).
10          $\mathcal{I}_i \leftarrow$ Algorithm 5.1 (Lawson flip).
11   **end**
12   **end**
13 **end**

Output: The communication network is locally a Delaunay triangulation after each sampling period.

---

are proved in [70] and [55], respectively (also cf. [116]). The following theorem summarises the results of this section.

Theorem 5.2 (Maintaining the Delaunay triangulation). *Consider a networked system with mobile agents P_i, $(i \in \mathcal{V})$ connected by the initial Delaunay triangulation communication graph $\mathcal{G}(0) = \mathcal{DT}\left(\mathcal{P}(0)\right) = \left(\mathcal{V}, \mathcal{E}(0)\right)$. If the sampling time T_s is chosen sufficiently small, the graph $\mathcal{G}(t)$, $(t > 0)$ equals a Delaunay triangulation after each sampling period in which all agents executed Algorithm 5.4 while they are moving continuously along their trajectories $p_i(t)$.*

Note that Algorithm 5.4 maintains the network locally from the perspective of an individual agent i so that its adjacent triangles are Delaunay. However, it is possible that the overall network is not a Delaunay triangulation temporarily after an edge flip due to Algorithm 5.1 because of the following reason. After a Lawson flip of the edge (i, j) by agent i, the agents i and j are no longer connected directly. Thus, an adjacent triangle of agent j might become non-Delaunay which is fixed the next time agent j executes Algorithm 5.4. Thus, if it is mandatory that the overall network equals a Delaunay triangulation after each sampling period, a solution is to execute Algorithm 5.4 multiple times after every sampling period by each agent.

Critical situations might occur if a single agent moves so far during one sampling period that it causes multiple topological events which has to be prevented by choosing an appropriate sampling time T_s. This choice and the number of topological events highly depend on the dynamical behaviour of the agents and the control task. Thus, it might be desirable to combine the proposed algorithm with event-based control techniques (cf. [92]) to adapt the sampling periods to the topological events. However, multiple topological events that occur simultaneously in different parts of the triangulation do not influence each other and, hence, do not pose any problem.

5.4 Summary and literature notes

This chapter presented the Delaunay triangulation as a communication graph for mobile multi-agent systems. A new condition to test the local Delaunay property of the Delaunay triangulation was stated in Theorem 5.1 and, based on that, a set of algorithms was developed that allows for maintaining the Delaunay triangulation while the agents are moving. By exploiting the properties of the Delaunay triangulation, these algorithms work with information that can be gathered by the current communication network and, hence, do not need a global coordinator. Furthermore, the proposed algorithms show that the Delaunay triangulation provides all necessary connections to maintain itself, i. e. no additional edges have to be introduced to the network for the sole sake of maintaining the Delaunay triangulation.

This chapter is based on the journal contribution [7] of the author, where the Algorithms 5.1–5.4 were developed. The proposed algorithms will be used in Chapter 6 to solve a vehicle formation control problem with guaranteed collision avoidance by exploitation of the distinctive properties of the Delaunay triangulation in combination with an appropriate controller.

6

Swarms of mobile agents

This chapter addresses the control of networked swarms of mobile agents. Formation control problems are considered that can be solved with a novel approach to the cooperative control of mobile agents with nonholonomic kinematics using artificial potential fields presented in Section 6.2. The main contribution of the author focuses on the collision avoidance with other agents. To this end, a way to guarantee collision avoidance that respects control input limitations using Morse potential functions is proposed in Section 6.3 and the results are extended to the case that the Delaunay triangulation is applied as the communication structure to the networked system in Section 6.4. It will be shown under which conditions the collision avoidance is preserved with the switching proximity communication network.

6.1 Introduction to swarming

Motivated by nature, the collective behaviour of flocks of birds or ants serves as a model for cooperative multi-agent systems with mobile agents. Formation control of networked vehicles has a wide area of applications to solve coverage, logistics or satellite control problems as well as in autonomous driving [32, 86, 106]. The fundamental common task in these applications is to create desired formations by adjusting the distance in between adjacent agents or by approaching individual destinations. In order to achieve scalability and reliability, the agents are connected by communication links represented by the edges of a network. The agents have to act based on information that can be gathered locally or by digital communication since there is no coordinating unit.

Networked multi-agent systems with the structure shown in Fig. 6.1 consist of N subsystems P_i which have a local controller C_i that is composed of a feedback unit F_i and a communication and decision unit D_i. Each subsystem P_i represents a mobile object that is located at the position

$$p_i(t) := \begin{pmatrix} x_i(t) \\ y_i(t) \end{pmatrix}, \quad i = 1, 2, \ldots, N$$

with the Cartesian coordinates $x_i(t)$ and $y_i(t)$. The agents are connected via a communication network that is represented by the graph $\mathcal{G} = (\mathcal{V}, \mathcal{E})$ with $\mathcal{V} = \{1, 2, \ldots, N\}$

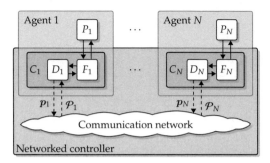

Fig. 6.1: Networked control structure of a multi-agent system

and $\mathcal{E}$ denoting the set of vertices and the set of edges, respectively. Two agents that are connected via an edge are called *neighbours*. The set of all neighbours of agent $i \in \mathcal{V}$ is $\mathcal{N}_i = \{j \mid (i, j) \in \mathcal{E}\}$ and the length of the corresponding edge $(i, j) \in \mathcal{E}$ is the distance

$$d_{ij}(t) := \left\| p_i(t) - p_j(t) \right\|, \quad i \neq j \tag{6.1}$$

between their positions.

In contrast to Chapter 5, this chapter will focus on the feedback unit F_i instead of the communication and decision unit D_i. The combination of the results on the communication network and the local controllers can be used to solve various control tasks.

- **Transition problems:** Mobile robots in a warehouse have to visit several destinations $p_i^\star$ in order to pick ordered goods. For that, the feedback F_i should guide all agents to their individual destination asymptotically:

$$\lim_{t \to \infty} \left\| p_i(t) - p_i^\star \right\| \overset{!}{=} 0, \quad i = 1, 2, \ldots, N. \tag{6.2}$$

- **Path tracking:** Lateral controllers have to be applied to vehicles in a platoon so that the controlled vehicles follow a prescribed path which is described by a connected one-dimensional manifold $\tilde{\mathcal{P}} \subset \mathbb{R}^2$. Given a projection $\tilde{p}_i(t) \in \tilde{\mathcal{P}}$ on the path, the feedback F_i should steer the agent towards the path:

$$\lim_{t \to \infty} \left\| p_i(t) - \tilde{p}_i(t) \right\| \overset{!}{=} 0, \quad i = 1, 2, \ldots, N. \tag{6.3}$$

- **Distance-based formations:** Flocks of mobile systems have to be established for cartography or coverage applications where an area has to be covered uniformly with sensors. Thus, each agent has to adjust its position $p_i(t)$ so that the distance $d_{ij}(t)$ to its neighbours $j \in \mathcal{N}_i$ equals the desired value $d^\star$:

$$\lim_{t \to \infty} d_{ij}(t) \overset{!}{=} d^\star, \quad (i, j) \in \mathcal{E}. \tag{6.4}$$

As in every control application with mobile agents, collision avoidance is a fundamental requirement that has to be achieved by the networked controller in Fig. 6.1. To this end, properties of the controlled agents induced by the local feedback F_i on the one hand and properties of the communication network on the other hand have to be found so that the combination of the controlled agents and the communication network guarantees collision avoidance. That is, during the transition from the initial positions $p_i(0)$, $(i = 1, 2, \ldots, N)$ to the desired position or formation (either of (6.2)–(6.4)), the distances should exceed the *minimum distance* d_0 at any time:

$$d_{ij}(t) \overset{!}{\geq} d_0, \quad t \geq 0, \quad i, j \in \mathcal{V}, \quad i \neq j. \tag{6.5}$$

The following sections will elaborate a control scheme with which these control aims can be achieved using artificial potential fields. To this end, this chapter will contribute with the following results.

1. A control scheme based on a gradient descent along an artificial potential field will be presented in Section 6.2.

2. In Section 6.3, it will be discussed how to choose an appropriate potential function in order to achieve a distance-based formation with guaranteed collision avoidance among connected agents under consideration of control input limitations.

3. The combination of a Delaunay network and the proposed controller will be analysed in Section 6.4 to find conditions under which the collision avoidance is extended among all agents of the networked system with the switching proximity network.

While the controller that will be presented in the next section is able to solve either of (6.2)–(6.4), this chapter will focus on distance-based formations due to the following reason. The transition problem and path tracking concern individual control aims of the agents and, hence, can be solved without a cooperation between multiple agents. Thus, to study the cooperative behaviour of the agents and the combination of local controllers with the choice of a communication structure, the distance-based formation (6.4) is considered. Nevertheless, the transition problem and path tracking will be addressed in Chapter 7 with further theoretical and experimental results.

6.2 Control of mobile agents with artificial potential fields

6.2.1 Geometric and kinematic description of a mobile agent

This chapter considers mobile agents that are described by nonholonomic kinematics. The geometrical description of such an agent is illustrated in Fig. 6.2. The orientation vector

$$e_i\big(\phi_i(t)\big) := \begin{pmatrix} \cos \phi_i(t) \\ \sin \phi_i(t) \end{pmatrix}, \quad \big\|e_i\big(\phi_i(t)\big)\big\| = 1 \tag{6.6}$$

with $\phi_i(t)$ denoting the orientation angle fixed to the body determines the direction of movement of the robot for positive velocity and the vector

$$n_i\big(\phi_i(t)\big) := \begin{pmatrix} -\sin\phi_i(t) \\ \cos\phi_i(t) \end{pmatrix}, \quad \big\|n_i\big(\phi_i(t)\big)\big\| = 1 \tag{6.7}$$

denotes the normal vector which is defined to satisfy

$$e_i^{\mathsf{T}}\big(\phi_i(t)\big)\, n_i\big(\phi_i(t)\big) = 0$$

so that $e_i\big(\phi_i(t)\big)$ and $n_i\big(\phi_i(t)\big)$ are orthogonal to each other.

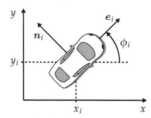

Fig. 6.2: Geometrical description of a mobile agent

The pose of an agent consists of the position $p_i(t)$ and the orientation angle $\phi_i(t)$ and evolves according to the nonholonomic kinematics

$$P_i : \begin{cases} \dot{p}_i(t) = e_i\big(\phi_i(t)\big)\, v_i(t) \\ \dot{\phi}_i(t) = \omega_i(t), \end{cases} \qquad \begin{pmatrix} p_i(0) \\ \phi_i(0) \end{pmatrix} = \begin{pmatrix} p_{i0} \\ \phi_{i0} \end{pmatrix} \tag{6.8}$$

with the control inputs $v_i(t)$ and $\omega_i(t)$ denoting the translational and rotational velocity, respectively.

The model (6.8) is a constrained integrator, i. e. the position $p_i(t)$ cannot evolve freely on a driving surface but rather specified by the orientation vector $e_i\big(\phi_i(t)\big)$. In order to approach any arbitrary position, the agent has to turn towards the destination with an appropriate rotational input $\omega_i(t)$ and, then, approach its destination with the translational input $v_i(t)$. The following sections will present a controller that solves this task simultaneously.

6.2.2 Artificial potential fields

The idea of the control with artificial potential fields is to introduce a multivariable function that penalises any deviation of the current overall configuration of the agents from the desired formation. To this end, consider the vector

$$p(t) = \big(p_1^{\mathsf{T}}(t) \quad p_2^{\mathsf{T}}(t) \quad \dots \quad p_N^{\mathsf{T}}(t)\big)^{\mathsf{T}}$$

of all concatenated positions. The overall potential field is defined as

$$P(p(t)) := \frac{1}{2} \sum_{i=1}^{N} \sum_{j \in \mathcal{N}_i} P_{ij}(d_{ij}(t)) \tag{6.9}$$

with the distance $d_{ij}(t)$ defined in (6.1). The overall potential field is composed of *relative potential functions* $P_{ij}(d_{ij}(t))$ that contribute to the overall potential $P(p(t))$ if the corresponding edge $(i, j) \in \mathcal{E}$ of the communication graph $\mathcal{G}$ is either larger or smaller than the desired value $d^\star$. They can be formally defined as follows.

Definition 6.1 (Relative potential function). A function $P_{ij}(d_{ij})$ is called a *relative potential function* if it possesses the following characteristics: In its domain $\mathcal{D}$, the function $P_{ij}(d_{ij})$ is nonnegative

$$P_{ij}(d_{ij}) : \mathcal{D} \mapsto \mathbb{R}^{\geq 0} \tag{6.10}$$

and continuously differentiable

$$P_{ij}(d_{ij}) \in C^1(\mathcal{D}). \tag{6.11}$$

Furthermore, it has the unique global minimum

$$P_{ij}(d^\star) = 0, \quad d^\star \in \mathcal{D}. \tag{6.12}$$

Functions that satisfy (6.10)–(6.12) have the consequence that $P(p(t))$ is equal to zero if and only if the control aim (6.4) is achieved. Consequently, the goal of the local feedback F_i of all agents $i \in \mathcal{V}$ is to minimise the overall potential by adjusting their position. This can be achieved with a continuous gradient descent along the overall potential. The following lemma shows how to determine the gradient from the perspective of an individual agent.

Lemma 6.1 (Local gradient of the overall potential fields). *The gradient of the overall potential field $P(p(t))$ from the perspective of an individual agent i is determined by*

$$\nabla_{p_i} P(p(t)) = \sum_{j \in \mathcal{N}_i} \frac{dP_{ij}(d_{ij}(t))}{dd_{ij}(t)} \frac{p_i(t) - p_j(t)}{d_{ij}(t)} \tag{6.13}$$

Proof. All addends in (6.9) that correspond to agents that are not connected to agent i do not contribute to the local gradient. With this observation, the gradient is determined as

$$\nabla_{p_i} P(p(t)) = \nabla_{p_i} \sum_{j \in \mathcal{N}_i} P_{ij}(d_{ij}(t)) = \sum_{j \in \mathcal{N}_i} \frac{dP_{ij}(d_{ij}(t))}{dd_{ij}(t)} \nabla_{p_i} d_{ij}(t) \tag{6.14}$$

by switching the gradient and the sum and by application of the chain rule. The gradient of the distance is given by

$$\nabla_{p_i} d_{ij}(t) = \nabla_{p_i} \|p_i(t) - p_j(t)\| = \frac{p_i(t) - p_j(t)}{\|p_i(t) - p_j(t)\|} = \frac{p_i(t) - p_j(t)}{d_{ij}(t)}, \qquad (6.15)$$

which yields the result (6.13) after insertion in (6.14). ∎

The negative gradient $-\nabla_{p_i} P(p(t))$ can be interpreted as the current desired direction of movement. It is a composition of the influences of all neighbours $j \in \mathcal{N}_i$ and, hence, it is possible that multiple neighbours cancel or reinforce each others influence on agent i. Furthermore, consider the following observation. The gradient of the distance $d_{ij}(t)$ is a unit vector due to (6.15). Thus, the direction of the movement is determined solely by $\nabla_{p_i} d_{ij}(t)$ since the derivative of the relative potential $P_{ij}(d_{ij}(t))$ with respect to the distance is a scalar. The magnitude of the influence of a neighbour on the other hand is determined by the choice of the slope of $P_{ij}(d_{ij}(t))$.

The following section will propose a method to choose the control inputs $v_i(t)$ and $\omega_i(t)$ of the mobile agent (6.8) so that the movement of the controlled agent follows the gradient (6.13) in negative direction.

6.2.3 A gradient-based control law

The negative gradient $-\nabla_{p_i} P(p(t))$ determined with Lemma 6.1 represents the desired direction of movement of an agent as in Fig. 6.3 (a). Since the agents are not aligned correctly to the gradient in general, the agents have to turn so that the orientation vector $e_i(\phi(t))$ and the gradient are parallel to each other. The approach of the controller to be proposed in this section is to decompose the control action in a translational and a rotational component that are executed with the control inputs $v_i(t)$ and $\omega_i(t)$. The translational part $v_i(t)$ is taken from the literature (cf. [77]) and the rotational part $\omega_i(t)$ is a novel approach based on the maximisation of the translational velocity.

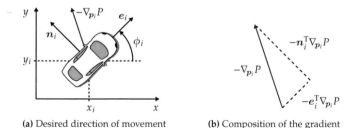

| (a) Desired direction of movement | (b) Composition of the gradient |

Fig. 6.3: Interpretation of the gradient descent

Translational velocity

The gradient vector $\nabla_{p_i} P(p(t))$ can be decomposed into a part that is parallel to $e_i(\phi(t))$ and a part that is parallel to $n_i(\phi(t))$ as shown in Fig. 6.3 (b) so that

$$\left\| \nabla_{p_i} P(p(t)) \right\|^2 = \left(e_i^T(\phi(t)) \nabla_{p_i} P(p(t)) \right)^2 + \left(n_i^T(\phi(t)) \nabla_{p_i} P(p(t)) \right)^2$$

holds true. The corresponding legs of the triangle can be interpreted as a decomposition of the control error. Now, suppose that the agent can only move along the orientation vector $e_i(\phi(t))$, i.e. $\omega_i(t)$ is fixed to be equal to zero in this consideration. Then, the proportional feedback of the error along the orientation vector according to

$$v_i(t) = -e_i^T(\phi_i(t)) \, \nabla_{p_i} P(p(t)) \tag{6.16}$$

makes the agent move along $e_i^T(\phi_i(t))$ until the error in the corresponding direction vanishes due to the integrator dynamics of (6.8) (cf. [77]):

$$\lim_{t \to \infty} e_i^T(\phi_i(t)) \, \nabla_{p_i} P(p(t)) = 0.$$

The feedback (6.16) with the gradient (6.13) has the following important property.

Lemma 6.2 (Nonincreasing potential). *For a set of subsystems P_i, $(i = 1, 2, \ldots, N)$ with the kinematics (6.8), the overall potential $P(p(t))$ is monotonically decreasing*

$$\dot{P}(p(t)) \le 0, \quad t \ge 0 \tag{6.17}$$

if all agents are actuated with the control law (6.16).

Proof. The property can be obtained straightforwardly by derivation of $P(p(t))$ as follows:

$$\dot{P}(p(t)) = \left(\nabla P(p(t)) \right)^T \dot{p}(t)$$

$$= \sum_{i=1}^{N} \left(\nabla_{p_i} P(p(t)) \right)^T \dot{p}_i(t)$$

$$= -\sum_{i=1}^{N} \left(e_i^T(\phi_i) \nabla_{p_i} P(p(t)) \right)^2$$

$$= -\left\| v(t) \right\|^2 \le 0$$

with $v(t) = \begin{pmatrix} v_1(t) & v_2(t) & \cdots & v_N(t) \end{pmatrix}^T$. ∎

Lemma 6.2 proves that the overall potential is monotonically decreasing. However, the overall potential field will not satisfy

$$\lim_{t \to \infty} P(p(t)) = 0$$

in general since the feedback (6.16) only works on the displacement along its current orientation $e_i(\phi(t))$. In the following, a feedback law for the rotational velocity $\omega_i(t)$ will be elaborated that steers the agent towards its desired direction of movement to solve this problem.

Control of the orientation

There are different approaches to the control of the orientation of agents with nonholonomic agents as in [50, 77, 94, 112, 113] for example. These publications have in common that the inverse tangent is applied for the control of the orientation of the robots. Especially in noisy experimental environments, the inverse tangent may lead to numerical issues since it is discontinuous. Furthermore, the proposed control laws for $\omega_i(t)$ in the aforementioned publications do not consider the potential field $P(p(t))$ but rather geometrical relations of the agents. This section proposes an alternative way to to control the orientation of a mobile agent with the kinematics (6.8) based on the potential field so that the closed-loop dynamics of each agent is solely determined by the choice of the relative potential functions $P_{ij}(d_{ij}(t))$.

The idea for the control of the rotational velocity $\omega_i(t)$ is to find the correct orientation by maximising the velocity with respect to the orientation angle $\phi_i(t)$ based on the following observation. The right-hand side of the translational velocity (6.16) is the inner product of two vectors (cf. Fig. 6.3). Its absolute value is at its maximum if both vectors are parallel to each other. Thus, the maximisation of the velocity by variation of the orientation angle $\phi_i(t)$ will turn the agent towards its desired direction of movement. Since the dynamical behaviour has integrating character according to the model (6.8), the maximisation can be done via the gradient ascent

$$\omega_i(t) = \mu\,\frac{\mathrm{d}v_i(t)}{\mathrm{d}\phi_i(t)} = -\mu\,\frac{\mathrm{d}e_i^{\mathrm{T}}(\phi_i(t))}{\mathrm{d}\phi_i(t)}\,\nabla_{p_i} P(p(t)) \tag{6.18}$$

with the learning rate μ. The derivative of the orientation vector with respect to the orientation angle is given by

$$\frac{\mathrm{d}e_i(\phi_i(t))}{\mathrm{d}\phi_i(t)} = n_i(\phi_i(t)),$$

which follows from the definitions (6.6) and (6.7). Insertion in (6.18) results in

$$\omega_i(t) = -\mu\,n_i^{\mathrm{T}}(\phi_i(t))\,\nabla_{p_i} P(p(t)), \tag{6.19}$$

which is very similar to (6.16) and allows for systematically generating control signals for a given potential function $P(p(t))$ without using the inverse tangent. Equation (6.19) will cause the robot to rotate as long as the gradient $\nabla_{p_i} P(p(t))$ and $n_i(\phi_i(t))$ are not orthogonal. The learning rate should be chosen as $\mu > 1$ in order to ensure that the dynamical behaviour of the orientation angle $\phi_i(t)$ settles down faster than the adjustment of the position due to the translational velocity (6.16).

Summary of the gradient-based controller

The combination of (6.16) and (6.19) yields the feedback unit

$$F_i : \begin{pmatrix} v_i(t) \\ \omega_i(t) \end{pmatrix} = - \begin{pmatrix} e_i^T(\phi_i(t)) \\ \mu\, n_i^T(\phi_i(t)) \end{pmatrix} \nabla_{p_i} P(p(t)) \qquad (6.20)$$

for some $\mu > 1$ which realises a descent along the gradient (6.13). The controller is characterised by the choice of the relative potential functions $P_{ij}(d_{ij}(t))$ which will be addressed in the next section.

Since the feedback unit (6.20) performs a gradient descent, the agents might get stuck in a local solution (*deadlock*) which happens when the partial derivative (6.13) is equal to zero due to the addends cancelling each other. The problem of getting stuck in an undesired equilibrium is well known in the literature on formation control (see [123] for an in-depth discussion). Furthermore, it has been shown that it is *not* possible to achieve global asymptotic stability with the potential function method since each other agent introduces at least one saddle point to the potential field [69]. However, local solutions generally occur due to sensitive symmetries which usually resolve due to measurement noise and, hence, occur rarely. Thus, convergence for almost all initial states is sufficient for most applications [50, 113].

Due to LaSalle's invariance principle, the agents converge to the largest invariant set of points $p_i(t), (i = 1, 2, \ldots, N)$ where $\dot{P}(p(t)) = 0$ holds true [66]. At this point, the derivatives of the relative potential functions with respect to the distances $d_{ij}(t), (i, j) \in \mathcal{E}$ are equal to zero. Consequently, the control inputs $v_i(t)$ and $\omega_i(t)$ generated by (6.20) are also equal to zero. Since the overall potential cannot increase as proven by Lemma 6.2, the positions $p_i(t), (i = 1, 2, \ldots, N)$ of the overall systems that eventually satisfy

$$\lim_{t \to \infty} \dot{P}(p(t)) = 0$$

are a stable equilibrium of the overall system dynamics. This consideration allows to conclude that the networked control system in Fig. 6.1 is stable with the feedback units (6.20). If and only if the overall potential satisfies

$$\lim_{t \to \infty} P(p(t)) = 0,$$

the agents approach the desired formation (6.4) exactly due to the properties of the relative potential functions $P_{ij}(d_{ij}(t))$ as per Definition 6.1.

6.3 Distance-based formations with guaranteed collision avoidance

6.3.1 Problem statement

The control aim is to reach the distance-based formation (6.4) as described in the introduction of this chapter. Due to the possibility of local solutions as discussed in the last
section, each agent $i \in \mathcal{V}$ has to adjust its position $p_i(t)$ so that the distance $d_{ij}(t)$ to its
neighbours $j \in \mathcal{N}_i$ equals the value $d^\star$ with a small tolerance $\epsilon > 0$:

$$\lim_{t \to \infty} \left| d_{ij}(t) - d^\star \right| \overset{!}{\leq} \epsilon, \quad (i,j) \in \mathcal{E}. \tag{6.21}$$

The value of ϵ depends on the network structure (cf. [105] for the notions of persistence
and rigidity). During the transition from the initial positions p_{i0}, $(i = 1, 2, \ldots, N)$ to the
desired distance-based formation (6.21), all distances should exceed a lower bound

$$d_{ij}(t) \overset{!}{\geq} d_0, \quad t \geq 0, \quad i, j \in \mathcal{V}, \quad i \neq j \tag{6.22}$$

to achieve collision avoidance (cf. (6.5)). Furthermore, the control aims (6.21) and (6.22) will
be considered under the circumstance that the control inputs are limited by the value $v_{\max}$:

$$\left| v_i(t) \right| \overset{!}{\leq} v_{\max}, \quad t \geq 0, \quad i \in \mathcal{V}. \tag{6.23}$$

Way of solution. This section will answer the question of how to choose the relative
potential functions $P_{ij}(d_{ij}(t))$ that achieve the control aims (6.21)–(6.23). To this end, the
following observation will be utilised: Due to its nonpositive derivative (6.17) provided
by Lemma 6.2, the maximum value of the overall potential field $P(p(t))$ is limited by the
initial value

$$P(p(t)) \leq P(p(0)).$$

Since the overall potential is the superposition of nonnegative (cf. (6.10)) relative potential
functions $P_{ij}(d_{ij}(t))$, it directly follows that

$$P_{ij}(d_{ij}(t)) \leq P(p(0)), \quad (i,j) \in \mathcal{E} \tag{6.24}$$

holds true. As a consequence, the range of possible values for the distance $d_{ij}(t)$ is limited
by (6.24). Thus, if the relative potential functions $P_{ij}(d_{ij}(t))$ are chosen so as to limit the
smallest possible value of $d_{ij}(t)$ to a value that is larger than the minimum distance d_0,
collision avoidance is guaranteed. A characterisation of the relative potential function
based on this idea will be elaborated in the following sections.

6.3.2 Collision avoidance with diverging potential functions

A common approach in the literature (cf. [38, 93, 94, 119, 135–137]) is to apply a relative potential function that is diverging due to a pole at $d_{ij}(t) = d_0$. The goal is to introduce a repulsion that is so large that it overweights all other effects from the neighbours $j \in \mathcal{N}_i$.

Example 6.1 (Diverging relative potential function). Consider the function

$$P_{ij}(d_{ij}) = \frac{d^\star - d_0}{d_{ij} - d_0} + \frac{d_{ij} - d^\star}{d^\star - d_0} - 1 \tag{6.25}$$

shown in Fig. 6.4. Along its domain $\mathcal{D} = (d_0, \infty)$, the function (6.25) is nonnegative, continuous and, furthermore, $P_{ij}(d^\star) = 0$ holds true so that it meets the requirements that are imposed on a relative potential function as per Definition 6.1.

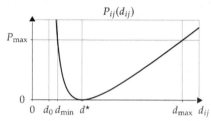

Fig. 6.4: Relative potential function (6.25) with a pole at d_0

For an initial value $P(p(0)) = P_{\max}$, the range of the argument d_{ij} is limited by the intersections of the function with the constant $P_{\max}$ due to (6.24) as illustrated in Fig. 6.4. For the relative potential function (6.25), the values of the maximum and minimum possible arguments are determined by

$$d_{\max/\min} = d^\star + \frac{d^\star - d_0}{2} \left(P_{\max} \pm \sqrt{P_{\max}(4 + P_{\max})} \right)$$

for given parameters d_0 and $d^\star$ that satisfy $d^\star > d_0$. $\qquad\square$

Due to the pole, the potential function (6.25) strives towards infinity if the argument approaches the minimum distance d_0 from above:

$$\lim_{d_{ij}(t) \downarrow d_0} P_{ij}(d_{ij}(t)) \to \infty. \tag{6.26}$$

Consequently, the control inputs $v_i(t)$ and $v_j(t)$ also increase indefinitely, which generates a repulsion between agents i and j of arbitrary magnitude, and collision avoidance of connected agents is hence achieved as shown by the following lemma.

> **Lemma 6.3** (Trivial collision avoidance). *Consider a set of subsystems $P_i, (i \in \mathcal{V})$ with the kinematics (6.8) controlled by the feedback (6.20) with a relative potential function $P_{ij}(d_{ij}(t))$ that satisfies (6.26). For any initial condition $p_i(0), (i \in \mathcal{V})$ that satisfies $d_{ij}(0) > d_0, (i,j) \in \mathcal{E}$, all distances of connected agents remain larger than the minimum distance d_0:*
>
> $$d_{ij}(t) > d_0, \quad t \geq 0, \quad (i,j) \in \mathcal{E}. \tag{6.27}$$

Proof. Since $P_{ij}(d_{ij}(t))$ is bounded by its initial value according to (6.24), it follows that the minimum length of each edge $(i,j) \in \mathcal{E}$ is bounded by a minimum value denoted by d_{min}:

$$d_{ij}(t) \geq d_{min}, \quad t \geq 0, \quad (i,j) \in \mathcal{E}.$$

Furthermore, due to the pole of $P_{ij}(d_{ij}(t))$ at d_0, it follows that $d_{min} > d_0$ holds true and, consequently, (6.27) follows which completes the proof. ∎

Lemma 6.3 shows that it is straightforward to prove collision avoidance among connected agents using diverging relative potential functions such as (6.25). However, the collision avoidance relies on the fact that the control input $v_i(t)$ gets indefinitely large if the distance $d_{ij}(t)$ approaches the minimum distance d_0 which will eventually violate (6.23).

In the next sections, a way to achieve collision avoidance under consideration of the control input limitation (6.23) will be presented. It uses the same strategy to narrow the distance to a left-closed interval so that the minimum possible distance is larger or equal than the minimum distance d_0. However, in contrast to the solution shown in this section, the following analysis will use continuous functions based on Morse potentials that do not have a pole. Thus, instead of increasing the repelling effect indefinitely, the attracting component is limited which yields a solution that respects the limitation (6.23).

6.3.3 Morse potential function

The Morse potential function is named after the physicist Philip M. Morse who introduced a specific potential function in the field of molecular physics to model the interatomic potential of diatomic molecules [101]. The potential function is defined as

$$P_{ij}(d_{ij}(t)) := k_1 \left(1 - e^{-k_2(d_{ij}(t) - d^\star)}\right)^2 \tag{6.28}$$

with nonnegative coefficients k_1, k_2. Its derivative with respect to the distance $d_{ij}(t)$ is

$$\frac{dP_{ij}(d_{ij}(t))}{dd_{ij}(t)} = 2k_1 k_2 \, e^{-k_2(d_{ij}(t) - d^\star)} \left(1 - e^{-k_2(d_{ij}(t) - d^\star)}\right). \tag{6.29}$$

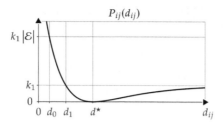

Fig. 6.5: Morse potential function (6.28)

The Morse potential function is nonnegative (6.10), continuous (6.11) and satisfies (6.12) so that it meets the requirements given in Definition 6.1 and, hence, it can be used as a relative potential function for the controller (6.20).

Figure 6.5 shows an example of a Morse potential function. For small distances $d_{ij}(t)$, it is shaped like a decaying exponential function and for large arguments, it stagnates and approaches the limit

$$\lim_{d_{ij} \to \infty} P_{ij}(d_{ij}) = k_1$$

asymptotically which is a major difference compared to the radially unbounded potential function (6.25) in Fig. 6.4. The distance d_1 denotes the argument for which the value of the function equals the coefficient k_1:

$$P_{ij}(d_1) = k_1. \tag{6.30}$$

Its value can be determined straightforwardly by insertion of (6.30) in the definition (6.28) which yields

$$d_1 = d^\star - \frac{\ln(2)}{k_2}. \tag{6.31}$$

Due to the shape of the Morse potential function, any distance $d_{ij}(t)$ that exceeds d_1 satisfies

$$d_{ij}(t) \in [d_1, \infty) \implies P_{ij}(d_{ij}(t)) \le k_1. \tag{6.32}$$

The implication (6.32) is a valuable property that will be used in the following section to determine the parameters k_1 and k_2 for which collision avoidance can be guaranteed for all initial conditions $p_{i0}, (i = 1, 2, \ldots, N)$ that satisfy

$$d_{ij}(0) \ge d_1, \quad (i, j) \in \mathcal{E}. \tag{6.33}$$

6.3.4 Parametrisation of Morse potential functions

The idea of this section is to satisfy the control aims (6.22) and (6.23) with a choice of appropriate parameters k_1 and k_2. For that, consider the worst-case scenario which can be described as follows. Suppose that all the energy represented by the overall potential field $P(p(t))$ in the networked system is spent on one critical edge $(p, q) \in \mathcal{E}$ to cause a collision intentionally. If the relative potential functions $P_{ij}(d_{ij}(t))$ are parameterised so that by conservation the maximum possible value of $P_{ij}(d_{ij}(t))$ is reached at the minimum distance d_0, then it is not possible to further decrease the distance $d_{pq}(t)$ of the critical edge (p, q) and collision avoidance is guaranteed. Furthermore, to account for the control input limitation, the repulsion in the worst-case scenario should respect the condition (6.23) by adjusting the slope of $P_{ij}(d_{ij}(t))$ at the minimum distance. In the following, the maximum value of the overall potential $P(p(t))$ is analysed to deduce conditions on the Morse potential function.

For initial conditions that satisfy (6.33), it follows that $P_{ij}(d_{ij}(t))$ is bounded by k_1 initially due to the property (6.32) which yields the following lemma concerning the overall potential.

Lemma 6.4 (Maximum potential). *Consider a set of N agents controlled by the feedback (6.20) with initial positions $p_{i0}, (i = 1, 2, \ldots, N)$ that satisfy (6.33). If the overall potential $P(p(t))$ is composed of Morse potential functions (6.28), the overall potential is bounded according to*

$$P(p(t)) \leq k_1 |\mathcal{E}|, \quad t \geq 0, \tag{6.34}$$

where $|\cdot|$ denotes the cardinality of a set, i. e. $|\mathcal{E}|$ is the number of edges in the networked control system.

Proof. Application of (6.32) to the definition of the overall potential (6.9) with the Morse potential function (6.28) yields

$$P(p(0)) \leq \frac{1}{2} \sum_{i=1}^{N} \sum_{j \in \mathcal{N}_i} k_1 = k_1 \underbrace{\frac{1}{2} \sum_{i=1}^{N} |\mathcal{N}_i|}_{2|\mathcal{E}|} = k_1 |\mathcal{E}|.$$

The bound holds true for $t > 0$ since $P(p(t))$ is monotonically decreasing due to Lemma 6.2 which proves the lemma. ∎

The design objectives are formalised as follows. The maximum possible value for $P_{ij}(d_{ij}(t))$ given by Lemma 6.4 should be reached at $d_{ij}(t) = d_0$ (cf. Fig. 6.5):

$$P_{ij}(d_0) \stackrel{!}{=} k_1 |\mathcal{E}|. \tag{6.35}$$

In order to take the input boundary (6.23) into account, the slope of the relative potential function given by the derivative (6.29) is chosen so as to satisfy

$$\frac{dP_{ij}(d_{ij}(t))}{dd_{ij}(t)}\bigg|_{d_{ij}(t)=d_0} \overset{!}{=} -v_{\max} \tag{6.36}$$

at the minimum distance d_0 so that the worst-case repulsion uses the maximum possible control action. The parameters that satisfy these design goals are given by the following theorem.

Theorem 6.1 (Parametrisation of the Morse potential). *Consider a set of subsystems P_i, $(i = 1, 2, \ldots, N)$ with the kinematics (6.8) controlled by the feedback (6.20) with Morse potential functions (6.28). For any initial condition p_{i0}, $(i = 1, 2, \ldots, N)$ that satisfies (6.33), the design objectives (6.35)–(6.36) are satisfied if the coefficients of the Morse potential function (6.28) are chosen according to*

$$k_1 = -\frac{(d_0 - d^\star) v_{\max}}{2 \ln\left(1 + \sqrt{|\mathcal{E}|}\right)\left(1 + \sqrt{|\mathcal{E}|}\right)\sqrt{|\mathcal{E}|}} \tag{6.37}$$

$$k_2 = -\frac{\ln\left(1 + \sqrt{|\mathcal{E}|}\right)}{d_0 - d^\star} \tag{6.38}$$

for a set of given parameters d_0, $d^\star$, $v_{\max}$ and a fixed network with the edge set $\mathcal{E}$.

Proof. The combination of (6.35) and the definition (6.28) of the Morse potential yields

$$k_1 \left(1 - e^{-k_2(d_0 - d^\star)}\right)^2 = k_1 |\mathcal{E}|$$
$$\Longleftrightarrow e^{-k_2(d_0 - d^\star)} = 1 \pm \sqrt{|\mathcal{E}|}. \tag{6.39}$$

Only the positive solution of the square root on the right-hand side of (6.39) is feasible since the exponential function is positive. By application of the natural logarithm, (6.38) follows after reformulation.

The coefficient k_1 is obtained by combining (6.36) with the derivative (6.29) to get

$$2k_1 k_2 e^{-k_2(d_0 - d^\star)} \left(1 - e^{-k_2(d_0 - d^\star)}\right) = -v_{\max}. \tag{6.40}$$

By application of (6.38), the exponential functions can be substituted according to

$$e^{-k_2(d_0 - d^\star)} = \left(1 + \sqrt{|\mathcal{E}|}\right).$$

Equation (6.40) then becomes

$$2\frac{k_1}{d_0 - d^\star} \ln\left(1 + \sqrt{|\mathcal{E}|}\right)\left(1 + \sqrt{|\mathcal{E}|}\right)\sqrt{|\mathcal{E}|} = -v_{max},$$

which results in (6.37) after reformulation. ∎

The coefficients of the relative potential functions $P_{ij}(d_{ij}(t))$ can be determined by (6.37) and (6.38) with which the control system is fully characterised. The values of $d_0, d^\star > d_0$ and $v_{max} > 0$ are free to choose or given by technical constraints. The number of edges $|\mathcal{E}|$ of the network on the other hand depends on the network structure that is used. The choice of an appropriate network structure will be addressed in Section 6.4.

6.3.5 Collision avoidance with Morse potential functions

With the parametrisation given by Theorem 6.1, collision avoidance can be guaranteed with the same reasoning as in Section 6.3.2: Since the maximum possible value of the overall potential field is bounded due to Lemma 6.4, it is not possible for an edge to become smaller than the minimum distance d_0.

Theorem 6.2 (Collision avoidance with neighbours). *Consider a set of subsystems P_i, $(i = 1, 2, \ldots, N)$ with the kinematics (6.8) controlled by the feedback (6.20) with Morse potential functions (6.28). For any two neighbours connected via the edge $(i, j) \in \mathcal{E}$, their distance is bounded according to*

$$d_{ij}(t) \in [d_0, \infty), \quad (i, j) \in \mathcal{E}, \quad t \geq 0$$

if the initial distances satisfy (6.33) and the parameters k_1 and k_2 are chosen according to Theorem 6.1.

Proof by contradiction. Suppose there is a collision, i.e. there is a pair of agents connected via the edge $(i, j) \in \mathcal{E}$ with a distance $d_{ij}(t)$ that is shorter than the minimum distance: $d_{ij}(t) < d_0$. Due to (6.35) established by Theorem 6.1 and the fact that $P_{ij}(d_{ij}(t))$ is monotonically decreasing for $d_{ij}(t) < d^\star$,

$$P_{ij}(d_{ij}(t)) > k_1 |\mathcal{E}|$$

holds true. However, this conclusion contradicts the maximum possible value of the overall potential given by Lemma 6.4. Thus, the hypothesis is wrong and, consequently, it is not possible for two neighbouring agents to collide. ∎

Discussion of the result

Theorem 6.2 proves collision avoidance for any two agents that are connected via the communication network $\mathcal{G}$. Thus, collision avoidance of the overall multi-agent system follows if a complete graph is applied. By introduction of

$$\gamma := \frac{\ln(2)}{\ln\left(1 + \sqrt{|\mathcal{E}|}\right)}, \tag{6.41}$$

the bound d_1 defined in (6.31) that determines all feasible initial distances (6.33), for which collision avoidance is guaranteed, can be reformulated by application of (6.38) as

$$d_1 = \gamma \, d_0 + (1 - \gamma) \, d^\star. \tag{6.42}$$

The result (6.42) is a convex combination due to the following observation. For a network with just one edge $|\mathcal{E}| = 1$, γ equals one whereas in a network with a large number $|\mathcal{E}|$ of edges, γ strives towards zero. The corresponding values of d_1 are d_0 and $d^\star$, respectively. Thus,

$$d_1 \in [d_0, d^\star] \tag{6.43}$$

holds true which can be interpreted as follows. Fewer edges in the communication graph $\mathcal{G}$ imply that d_1 is closer to d_0 and, thus, the feasibility region (6.33) becomes larger. The solution is hence less conservative. Conversely, a graph with a large number of edges requires d_1 to be closer to $d^\star$ which results in a smaller feasibility region (6.33).

The selection of an appropriate communication graph $\mathcal{G}$ has to satisfy two contradicting aims. First, the observation (6.42)–(6.43) shows that with the parametrisation proposed by Theorem 6.1, it is desirable to have few edges in the network. Second, the graph $\mathcal{G}$ has to be dense enough to provide a sufficient number of edges to guarantee collision avoidance among all agents. As a trade-off, the Delaunay triangulation as a switching communication graph that approximates the complete graph (cf. [65]) is proposed in the next section which will be shown to guarantee collision avoidance in the overall networked control system with less communication links than the complete graph.

6.4 Combination of Morse potential functions with Delaunay networks

6.4.1 The Delaunay triangulation as a communication network

The Delaunay triangulation has been introduced in Chapter 5 where distributed algorithms were developed with which the agents are able to maintain the Delaunay triangulation network while they are moving. This section proposes to use the Delaunay triangulation as the communication network structure for the networked control system to achieve the distance-based formation (6.21).

The discussion of the last section showed that the number of edges plays an important role in the application of the proposed controller parametrisation according to Theorem 6.1. By application of the Delaunay triangulation as the communication network, the structure changes at discrete point in time and, thus, $\mathcal{E}(t)$ is time-dependent. A Delaunay triangulation with N vertices has

$$\left|\mathcal{E}(t)\right| = 3\,(N - 1) - M(t)$$

edges where

$$M(t) \in [3, N]$$

denotes the number of agents on the boundary of the convex hull of the set of all agent positions $\mathcal{P}(t) = \{p_1(t), p_2(t), \dots, p_N(t)\}$ (cf. Section 5.1.2 and [45, 72]). Note that $M(t)$ is also time-dependent due to the movement of the agents and, hence, the number of agents on the convex hull may vary. Consequently, the Delaunay triangulation has a minimum of $2N-3$ edges if $M(t) = N$ and a maximum of $3(N-2)$ edges if $M(t) = 3$:

$$\left|\mathcal{E}(t)\right| \in [2N - 3,\ 3(N - 2)]. \tag{6.44}$$

The number of edges of a Delaunay triangulation increases linearly with the number N of agents whereas the complete graph has $N(N - 1)/2$ edges which is quadratic in N. Thus, the Delaunay triangulation has significantly less communication links especially for networked systems with a large number of agents.

6.4.2 Properties of the Lawson flip

The collision avoidance for neighbours connected via a fixed communication graph provided by Theorem 6.2 is based on the limitation (6.34) of the overall potential given by Lemma 6.4. During the transition of the agents from their initial positions p_{i0} to the desired formation (6.21), the Delaunay triangulation is maintained with Algorithm 5.4 developed in Chapter 5. Thus, there are edges that are removed and introduced to the network whenever a topological event occurs. Since edges that are replaced due to the Lawson flip are not generally of the same length, the overall potential $P\big(p(t)\big)$ jumps at the time instances of topological events and is, therefore, discontinuous. Critical situations might occur if a newly established edge is significantly shorter than the edge that the new one replaces. Due to the shape of the Morse potential in Fig. 6.5, a very short edge might contribute too much to the overall potential field $P\big(p(t)\big)$ so that the bound (6.34) is not respected. Thus, in order to investigate whether a jumping overall potential $P\big(p(t)\big)$ jeopardises the collision avoidance of the overall system, it has to be determined under

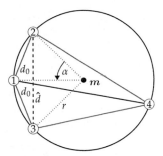

Fig. 6.6: The long edge $(1,4)$ will be replaced by the short edge $(2,3)$ due to the Lawson flip

which condition the contribution of a new edge still respects the upper bound of the overall potential $P(p(t))$ given by Lemma 6.4.

To this end, consider Fig. 6.6 which illustrates a convex quadrilateral. If agent $i = 1$ moves to the left in the present configuration, the edge $(1,4)$ will be replaced by the edge $(2,3)$ due to the Lawson flip (cf. Section 5.3.1). Let $\hat{d}$ denote the length of the new edge $(2,3)$. The following lemma determines a lower bound of $\hat{d}$ under the assumption that the length of both outer edges $(1,2)$ and $(1,3)$ is equal to the minimum distance d_0.

Lemma 6.5 (Shortest possible edge). *Consider a set of four agents that form a convex quadrilateral as in Fig. 6.6. The shortest possible length of an edge that is established due to the Lawson flip is given by*

$$\hat{d}_{\min} = \sqrt{2}\, d_0 \qquad (6.45)$$

if the current configuration of the agents does not contain a collision.

Proof. The law of cosines yields

$$\hat{d} = r\, \sqrt{2(1 - \cos(2\alpha))} = 2r\, \sin(\alpha) \qquad (6.46)$$

after application of the identity $1 - \cos(2\alpha) = 2\sin^2(\alpha)$ with r and α denoting the radius of the circumcenter and the angle spanned by the vertices of the Delaunay triangulation and the circumcenter m in Fig. 6.6, respectively. Similarly, the expression

$$d_0 = r\, \sqrt{2(1 - \cos(\alpha))}$$

is obtained with the law of cosines which is equivalent to

$$\alpha = \arccos\left(1 - \frac{d_0^2}{2r^2}\right). \qquad (6.47)$$

Insertion of (6.47) in (6.46) results in

$$\hat{d} = d_0 \sqrt{4 - \frac{d_0^2}{r^2}} \tag{6.48}$$

using the identity $\sin(\arccos(x)) = \sqrt{(1-x)(1+x)}$.

The densest possible convex quadrilateral that does not contain a collision has outer edges of the length d_0 as illustrated in Fig. 6.7. Thus, the smallest possible radius r follows from Pythagoras' theorem as $r = d_0/\sqrt{2}$ which results in the image range

$$\hat{d} \in [\sqrt{2}\,d_0, 2d_0]$$

of (6.48) given the domain $r \in [d_0/\sqrt{2}, \infty)$ of (6.48) which proves the lemma. ∎

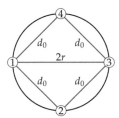

Fig. 6.7: Minimum radius of a circumcircle through four cocircular points without collision

Note that in addition to the edges that are established due to the Lawson flip, there are edges that are established due to the maintenance of the convex hull with Algorithm 5.3. Whenever an agent enters the interior of the convex hull of $\mathcal{P}(t)$, a new edge on the convex hull is established that also causes the overall potential field $P(\boldsymbol{p}(t))$ to jump. However, the smallest possible length of such an edge is given by $2d_0$ (cf. Fig. 5.10 (b)). Thus, if an edge established by the Lawson flip with the critical length (6.45) does not cause a collision as will be elaborated in the following, a new edge with the length $2d_0$ established by Algorithm 5.3 will also not cause a collision.

6.4.3 Collision avoidance with the Delaunay triangulation

Lemma 6.5 helps to guarantee collision avoidance as follows. The range of $d_{ij}(t)$ in (6.32) characterised by (6.42) for which the Morse potential respects the bound $P_{ij}(d_{ij}(t)) \leq k_1$ has to include the value $\hat{d}_{\min}$ provided by (6.45) so that a new edge that is established by the Lawson flip also respects the bound of the relative potential. Thus, the condition

$$d_1 \overset{!}{\leq} \hat{d}_{\min} \tag{6.49}$$

has to be satisfied so that Lemma 6.4 also holds for the switching Delaunay triangulation when the agents are moving. Application of the results (6.42) and (6.45) to the condition (6.49) results in

$$d^\star \overset{!}{\leq} \frac{\sqrt{2} - \gamma}{1 - \gamma} d_0 \tag{6.50}$$

with γ defined in (6.41). Under condition (6.50), collision avoidance of the networked control system is guaranteed with the Delaunay triangulation as shown by the following theorem.

Theorem 6.3 (Collision avoidance with the Delaunay triangulation). *Consider a set of subsystems $P_i, (i \in \mathcal{V})$ with kinematics (6.8) and initial distances that satisfy (6.33). Suppose the agents are connected via the Delaunay triangulation $\mathcal{DT}(\mathcal{P}(t)) = (\mathcal{V}, \mathcal{E}(t))$ and controlled by (6.20) with Morse potential functions (6.28) parameterised by Theorem 6.1. The distance in-between any two agents is bounded according to*

$$d_{ij}(t) \in [d_0, \infty), \quad t \geq 0, \quad i, j \in \mathcal{V}, \quad i \neq j$$

if the desired distance $d^\star$ is chosen so as to meet (6.50).

Proof. The condition (6.50) ensures

$$\hat{d}_{\min} \in [d_1, \infty)$$

so that the contribution of a newly established edge to the overall potential $P(p(t))$ does not exceed k_1 due to (6.32). Thus, the overall potential respects the bound

$$P(p(t)) \leq 3(N - 2) k_1, \tag{6.51}$$

which extends the result of Lemma 6.4 using the upper limit of the number of edges (6.44) of the Delaunay triangulation. By choosing

$$k_1 = -\frac{(d_0 - d^\star) v_{\max}}{2 \ln(1 + \xi)(1 + \xi)\xi} \tag{6.52}$$

$$k_2 = -\frac{\ln(1 + \xi)}{d_0 - d^\star} \tag{6.53}$$

according to Theorem 6.1 with $\xi = \sqrt{3(N - 2)}$, collision avoidance follows with the same reasoning as in Theorem 6.2 since an edge that is shorter than d_0 would contradict the maximum possible value (6.51) of the overall potential. ∎

Theorem 6.3 shows that collision avoidance between all agents in the networked control system can be achieved with significantly less communication links than the complete graph by application of the Delaunay triangulation. This important result is obtained due to the property of the Delaunay triangulation that the closest neighbours are always connected (Lemma 5.2). As it has been shown in Chapter 5 that the Delaunay triangulation can be maintained with decentralised algorithms that are executed by the agents, the proposed method guarantees collision avoidance without a coordinating unit in real time for any initial positions p_{i0}, $(i = 1, 2, \ldots, N)$ that satisfy (6.33).

6.4.4 Summary of the procedure

The proposed solution to the distance-based formation problem with guaranteed collision avoidance is summarised in Algorithm 6.1. The minimum distance d_0 and the control input limitation v_{max} are supposed to be known from the technical set-up. The initial Delaunay triangulation has to be constructed with appropriate algorithms (cf. Section 5.3.2).

Algorithm 6.1 (Collision-free formation control).

Input: Minimum distance d_0, control input limitation v_{max}, number N of agents with kinematics (6.8) and initial Delaunay triangulation $\mathcal{DT}(\mathcal{P}(0)) = (\mathcal{V}, \mathcal{E}(0))$.

1 Maintain the Delaunay triangulation with Algorithm 5.4.
2 Choose the desired distance $d^\star$ according to (6.50) with

$$\gamma = \frac{\ln(2)}{\ln\left(1 + \sqrt{3(N-2)}\right)}.$$

3 Determine k_1 and k_2 with (6.52) and (6.53), respectively.
4 Use the networked controller given by (6.20) with the gradient (6.13) and the derivative of the Morse potential function (6.29).

Output: Networked control system (Fig. 6.1) with mobile subsystems P_i that achieves the control aims (6.21)–(6.23).

6.5 Summary and literature notes

This chapter addressed the collision avoidance of networked mobile agents that should form a distance-based formation. While the collision avoidance in the relevant literature is usually achieved with unbounded control inputs, Theorem 6.1 proposes a way to parametrise Morse potential functions that guarantee collision avoidance under consideration of control input limitations as shown by Theorem 6.2. The results have been extended to the case that a switching proximity network in form of a Delaunay triangulation is applied to the networked control system and it has been shown by Theorem 6.3 that collision avoidance among all agents in the networked system is achieved with Algorithm 6.1.

This chapter is based on the following contributions of the author. The controller (6.20) has been developed in [1] in order to solve the transition problem that will be addressed among other experiments in the following chapter. The main results of this chapter given by Theorem 6.1–6.3 have been published in [11].

Practical implementation and experimental evaluation 7

This chapter presents several experimental studies that evaluate the theoretical results of the previous chapters. To this end, the experimental plant, with which the measurements are conducted, is presented in Section 7.1. The experiments to be discussed are path tracking, vehicle platooning, distance-based formations and the transition problem given in Sections 7.2–7.5, respectively. This chapter will introduce auxiliary controllers that form the basis for the implementation of the control methods proposed in this thesis in order to verify the effectiveness of the methods in a laboratory environment.

7.1 SAMS: Synchronisation of autonomous mobile systems

7.1.1 Set-up of the laboratory plant

The experimental plant SAMS (*synchronisation of autonomous mobile systems*) is used at the Institute of Automation and Computer Control at the Ruhr-University Bochum to test methods for the coordination of multi-agent systems (Fig. 7.1). This section provides an overview of the set-up. The interested reader is referred to [145] for more details.

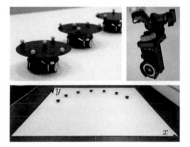

Fig. 7.1: SAMS with the robots (top left), camera (top right) and the driving surface with Cartesian coordinates (bottom) [145]

The experimental plant consists of a 4 m by 4 m large driving surface on which up to twelve mobile robots can move. Their positions are tracked by a camera system and the control inputs are transmitted to the robots via a radio module from a central computer. In the following, the components of the plant are explained briefly.

- **Robots:** The mobile robots m3pi of the manufacturer Pololu can move on the driving surface with their differential drive that consists of two independent wheels each driven by a DC motor. The control inputs are received via an XBee radio module. A plate with five reflecting marker balls, which are arranged in an individual pattern on each robot, allows each robot to be identified uniquely by the camera system.

- **Camera system:** Twelve Flex 3 infra red cameras of the company OptiTrack arranged evenly around the driving surface determine both the position $p_i(t)$ and the orientation angle $\phi_i(t)$ of the robots. This data is transmitted to the computer via a serial interface.

- **Computer:** The software Motive running on the computer evaluates the image data of the camera system in real time and transmits it to Matlab. Using the proposed control algorithms, the control inputs $u_i(t)$ and $\omega_i(t)$ of the robots are calculated based on the measured variables.

- **XBee module:** The control inputs are transmitted wirelessly to the robots via an XBee radio module. An individual ID identifier indicates for which robot the transmitted control input is determined.

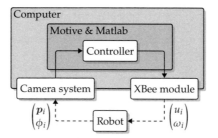

Fig. 7.2: Schematic block diagram of a robot control loop

Figure 7.2 shows the schematic structure of the control loop of a robot. Both the camera system and the XBee module have hardware and software components and represent part of the measuring and actuating elements of the control loop, respectively. All controllers are implemented as discrete-time algorithms in Matlab with a sample time of 30 ms. There are no significant transmission delays or packet losses in the experimental set-up used here. Note that even though the experimental set-up has a centralised computer for convenience, all methods are designed to be implemented locally on the robots in a distributed manner.

7.1.2 Mathematical description of a robot

The model that can be applied to describe a robot has been introduced in the last chapter in Section 6.2.1 and is summarised briefly in the following. The experiments in this chapter will be carried out with a set of N identical robots with a differential drive. The vector

$$p_i(t) = \begin{pmatrix} x_i(t) \\ y_i(t) \end{pmatrix}$$

denotes the position of a robot in Cartesian coordinates and the orientation angle fixed to the body is denoted by $\phi_i(t)$.

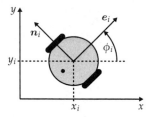

Fig. 7.3: Geometrical description of a robot with differential drive

The geometrical description of a robot is illustrated in Fig. 7.3. The orientation vector

$$e_i(\phi_i(t)) = \begin{pmatrix} \cos\phi_i(t) \\ \sin\phi_i(t) \end{pmatrix}$$

determines the direction of movement of the robot for positive velocity and the vector

$$n_i(\phi_i(t)) = \begin{pmatrix} -\sin\phi_i(t) \\ \cos\phi_i(t) \end{pmatrix} \tag{7.1}$$

denotes the normal vector which is orthogonal to $e_i(\phi_i(t))$.

The pose of a robot is determined by the position $p_i(t)$ and the orientation $\phi_i(t)$. The position evolves according to

$$\dot{p}_i(t) = e_i(\phi_i(t))\, u_i(t) \tag{7.2}$$

with the control input $u_i(t)$ that describes the *commanded* translational velocity which shows that the movement of a robot is always directed towards the orientation vector. The orientation angle is determined by the integrator dynamics

$$\dot{\phi}_i(t) = \omega_i(t) \tag{7.3}$$

with $\omega_i(t)$ denoting the rotational velocity which is the second input of the robot. The combination of (7.2) and (7.3) yields the state-space representation

$$R_i : \begin{cases} \dot{p}_i(t) = e_i\big(\phi_i(t)\big)\, u_i(t) \\ \dot{\phi}_i(t) = \omega_i(t), \end{cases} \qquad \begin{pmatrix} p_i(0) \\ \phi_i(0) \end{pmatrix} = \begin{pmatrix} p_{i0} \\ \phi_{i0} \end{pmatrix} \qquad (7.4)$$

of a robot with nonholonomic kinematics. The control inputs $u_i(t)$ and $\omega_i(t)$ are induced by the technical control inputs which are the left wheel and the right wheel voltages. The dynamical behaviour of the DC motors is assumed to be very quick and, hence, omitted in the model (7.4). However, in contrast to the similar model (6.8) that has been used in the last section, the model (7.4) uses the *commanded* velocity $u_i(t)$ instead of $v_i(t)$ which denotes the actual velocity. With this distinction, a possible lag between $u_i(t)$ and $v_i(t)$ can be considered in the controller design process which will be discussed in the respective sections in the following (cf. model (7.41)).

7.1.3 List of experiments

The experimental plant SAMS will be used to study the following classes of control problems concerning the cooperative control of networked vehicles:

1. **Path tracking:** For a given path, a lateral controller should steer the robot towards the path for any commanded velocity. The lateral controller is a low-level controller that complements a high-level longitudinal controller that is applied for platooning as an example. The path tracking control problem will be discussed and solved in Section 7.2.

2. **Vehicle platooning:** Longitudinal controllers as ACC are the counterpart of lateral controllers that should achieve cooperative behaviour between multiple autonomous vehicles. Section 7.3 will experimentally evaluate the theoretical results of Chapter 4 with platoons of robots. Furthermore, the results are extended to the case that multiple platoons have to merge in order to pass a lane reduction.

3. **Distance-based formations:** In contrast to the combination of lateral and longitudinal controllers, formation control problems concern the cooperative control of multiple mobile agents on a plane without prescribed paths. In Section 7.4, an experimental evaluation of the results of Chapters 5 and 6 is presented to study the combination of local controllers with a switching Delaunay triangulation network structure.

4. **Transition problem:** The gradient-based control scheme presented in Chapter 6 will be extended so that the artificial potential field also considers the deviation of an agent from its destination. The result presented in Section 7.5 is a controller that performs a transition from the initial position of a robot to its desired destination and, furthermore, that avoids other robots that are too close to achieve a collision-free transition.

The goal of the experiments is to evaluate the effectiveness of the methods presented in the previous chapters in an experimental environment. In addition to that, the following experiments also serve as an outlook by pointing out fields of application and it will be discussed what the measurements show beyond the theoretical results.

7.2 Path tracking control

7.2.1 Introduction to path tracking

This section proposes a lateral controller which should guide a mobile system to a prescribed path. In the context of autonomous driving, lateral control or path tracking describes the problem of keeping a mobile system on its lane or changing to an adjacent lane for merging or overtaking manoeuvres. The lateral controller complements the longitudinal controller which is used for distance control or cooperative intersection management [22].

Unfortunately, the notion of path tracking is not consistently used in literature and is often confused with trajectory tracking. While the latter describes the problem of navigating a mobile system along a continuous series of positions at very specific points in time, a lateral controller should navigate a mobile system along a prescribed path without imposing any restriction on the time. In other words, trajectory tracking fixes the velocity while path tracking leaves the velocity free to choose. Thus, path tracking allows for implementing high-level cooperative control structures for longitudinal control problems as, for example, platooning, obstacle avoidance or merging.

Path tracking control techniques usually work in two steps. First, the desired position on the path is determined by a projection of the current position and, second, an appropriate lateral controller compensates the deviation of the mobile system from this desired position. In [23], it has been shown that projection-based control techniques are able to reduce model complexity. Thus, lateral controllers allow for describing mobile systems by simplified models along the desired path which can be used to find appropriate longitudinal controllers.

Control aim

The control aim of path tracking is to guide a robot R_i to a prescribed path and keep it on the path by choosing $\omega_i(t)$. Thus, the lateral controller should steer the robot for any nonnegative commanded velocity $u_i(t)$, which is dictated by a high-level longitudinal controller, towards a desired path denoted by the set $\tilde{\mathcal{P}} \subset \mathbb{R}^2$. Formally, the path $\tilde{\mathcal{P}}$ is a connected one-dimensional manifold, e. g.

$$\tilde{\mathcal{P}} = \left\{ \begin{pmatrix} \tilde{x} \\ \tilde{y} \end{pmatrix} \mid \tilde{x}^2 + \tilde{y}^2 = R^2 \right\}$$

for a circular path with radius R. The desired Cartesian position on the path

$$\tilde{p}_i(t) \in \tilde{\mathcal{P}}$$

is obtained by orthogonal projection which satisfies

$$\left(\tilde{p}_i(t) - p_i(t) \right)^{\mathsf{T}} e_i \left(\tilde{\phi}_i(t) \right) = 0. \tag{7.5}$$

Equation (7.5) means that the displacement of the robot with respect to the path is orthogonal to the desired direction of movement at any time (cf. Fig. 7.5). The desired orientation angle $\tilde{\phi}(t)$ is determined by the slope of the tangent of the path at the position $\tilde{p}(t)$. It is assumed that the path is constructed so that $\tilde{p}(t)$ is continuously differentiable, i. e. the path is smooth.

Problem 7.1 (Path tracking control aim). Given a robot (7.4) and a path $\tilde{\mathcal{P}}$, find a rotational velocity $\omega_i(t)$ that steers the robot towards the path so that

$$\lim_{t \to \infty} \left\| \tilde{p}_i(t) - p_i(t) \right\| = 0 \tag{7.6}$$

$$\lim_{t \to \infty} \left\| \tilde{\phi}_i(t) - \phi_i(t) \right\| = 0 \tag{7.7}$$

hold true for any arbitrary $u_i(t) \geq 0$.

Note that a controller that solves Problem 7.1 does not perform trajectory tracking since the commanded velocity $u_i(t)$ is not prescribed. Furthermore, path tracking is *not* a cooperative task, i. e. for a set of N robots, each robot concerns Problem 7.1 on its own. Thus, for the sake of readability, the individual index i is omitted in the rest of this section.

7.2.2 A gradient-based path tracking controller

In order to achieve the control aims (7.6) and (7.7), the error signals

$$e_{\mathrm{p}}(t) := \tilde{p}(t) - p(t) \tag{7.8}$$

$$e_{\phi}(t) := \tilde{\phi}(t) - \phi(t) \tag{7.9}$$

are introduced so that Problem 7.1 is achieved if and only if the error signals converge to zero asymptotically. The lateral controller to be proposed is composed of two parts. The first part should adjust the orientation of the robot so that the error (7.9) vanishes. The second part should steer the robot towards the path if the error (7.8) is not equal to zero.

The approach of the first component is as follows. Suppose that the robot starts on the path, i.e. $e_p(0) = 0$. The lateral controller should keep the robot on the path by an appropriate rotational input. For that, consider the feedback

$$\omega_1(t) = \dot{\tilde{\phi}}(t) + k_1\, e_\phi(t). \tag{7.10}$$

The combination of the second state equation (7.4) with $\omega_1(t)$ yields

$$\dot{\phi}(t) = \dot{\tilde{\phi}}(t) + k_1\, e_\phi(t)$$
$$\Longleftrightarrow \dot{e}_\phi(t) = -k_1\, e_\phi(t),$$

which shows that any error concerning the orientation vanishes asymptotically for a positive k_1 (cf. [124]).

The second part will be developed with a gradient-based approach with the results of Section 6.2. To this end, consider the quadratic potential function

$$P\left(e_p(t)\right) = \frac{1}{2}\left\|e_p(t)\right\|^2.$$

Its gradient is determined according to

$$\nabla P\left(e_p(t)\right) = \frac{dP\left(e_p(t)\right)}{d\left\|e_p(t)\right\|}\, \nabla\left\|e_p(t)\right\| = \left\|e_p(t)\right\| \frac{e_p(t)}{\left\|e_p(t)\right\|} = e_p(t).$$

Note that the lateral controller should only steer the robot, i.e. it works with the rotational velocity $\omega(t)$ only. Hence, by application of the gradient descent (6.19) and the substitution $k_2 := -\mu$, the second component of the lateral controller is given by

$$\omega_2(t) = k_2\, e_p^{\mathsf{T}}(t)\, n\big(\phi(t)\big). \tag{7.11}$$

The proposed lateral controller is the superposition of (7.10) and (7.11) given by

$$\omega(t) = \omega_1(t) + \omega_2(t)$$
$$= \dot{\tilde{\phi}}(t) + k_1\, e_\phi(t) + k_2\, e_p^{\mathsf{T}}(t)\, n\big(\phi(t)\big). \tag{7.12}$$

In the following, it will be shown that the lateral controller (7.12) solves Problem 7.1 for appropriately chosen parameters k_1 and k_2. To this end, the next section elaborates on the relation between the robot state (cf. (7.4)) and the error signals and derives an error model that will be used to analyse the behaviour of the robot with the input (7.12).

The structure induced by the proposed controller (7.12) is shown in Fig. 7.4. The controller works in two steps. First, the projection unit uses the current measurements of $p(t)$ and $\phi(t)$ to determine the projection $\tilde{p}(t)$, $\tilde{\phi}(t)$ as well as the derivative $\dot{\tilde{\phi}}(t)$ and, furthermore, the translational velocity $v(t)$ as the new output. With the projected quantities, the path controller then determines the rotational velocity $\omega(t)$ that should steer the robot towards the path. The result is the new system Σ that represents the path-controlled robot which takes the commanded velocity $u(t)$ as the input and yields the translational velocity $v(t)$ as the output.

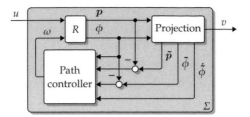

Fig. 7.4: Schematic path tracking control loop

Error dynamics

The goal of this section is to derive an error model that describes the dynamical behaviour of $e_p(t)$ and $e_\phi(t)$. Locally, the path $\tilde{\mathcal{P}}$ subdivides the driving surface into two regions. Thus, two cases have to be considered in the following. To this end, let the symbol $\angle$ denote the angle spanned by two position vectors with the same base.

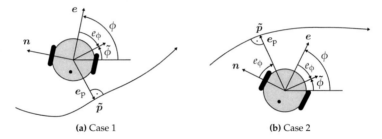

(a) Case 1   **(b)** Case 2

Fig. 7.5: Angles of the orthogonal projection

Case 1. The robot is on the left of the path in travelling direction as shown in Fig. 7.5 (a). The angles spanned by the vectors are given by

$$\angle e, e_p = e_\phi - \pi/2, \tag{7.13}$$

$$\angle n, e_p = e_\phi - \pi. \tag{7.14}$$

Case 2. In the situation shown in Fig. 7.5 (b), the robot is on the right of the path in travelling direction. The angles spanned by the vectors are given by

$$\angle e, e_p = e_\phi + \pi/2, \tag{7.15}$$

$$\angle n, e_p = e_\phi. \tag{7.16}$$

Note that the angles are shifted by π between the two cases because of the following reason. The direction of $e_p(t)$ switches to the opposite direction if the robot oversteps the path. However, all other quantities remain consistent.

The two cases will be used to find a common description of the control error of the lateral controller as follows. The combination of (7.4) with the lateral controller (7.12) yields

$$\dot{e}_\phi(t) = -k_1\, e_\phi(t) - k_2\, e_p^T(t)\, n\big(\phi(t)\big). \tag{7.17}$$

The scalar product in the second addend can be written as

$$
\begin{aligned}
e_p^T(t)\, n\big(\phi(t)\big) &= \big\|e_p(t)\big\|\ \underbrace{\big\|n\big(\phi(t)\big)\big\|}_{=\,1,\ \text{cf. (7.1)}}\ \cos\big(\angle n, e_p\big) \\
&= \begin{cases} -\big\|e_p(t)\big\| \cos\big(e_\phi(t)\big), & \text{case 1} \\ \big\|e_p(t)\big\| \cos\big(e_\phi(t)\big), & \text{case 2} \end{cases}
\end{aligned}
\tag{7.18}
$$

using (7.14) and (7.16). By introduction of the new scalar error variable

$$z(t) := \mp\big\|e_p(t)\big\| = \begin{cases} -\big\|e_p(t)\big\|, & \text{case 1} \\ \big\|e_p(t)\big\|, & \text{case 2,} \end{cases} \tag{7.19}$$

which can be interpreted as the signed deviation of the robot from the path, (7.18) can be written in closed form as

$$e_p^T(t)\, n\big(\phi(t)\big) = z(t) \cos\big(e_\phi(t)\big). \tag{7.20}$$

Given this preliminary work, the error dynamics is described by the following lemma.

Lemma 7.1 (Path tracking error dynamics). *Consider a robot (7.4) that is steered by the lateral controller (7.12). The dynamical behaviour of the error signals (7.8) and (7.9) is determined by the state-space model*

$$\Sigma_e : \begin{cases} \dot{z}(t) = \sin\big(e_\phi(t)\big)\, u(t) \\ \dot{e}_\phi(t) = -k_1\, e_\phi(t) - k_2\, z(t) \cos\big(e_\phi(t)\big). \end{cases} \tag{7.21}$$

Proof. With (7.20), the orientation error state equation (7.17) is given by

$$\dot{e}_\phi(t) = -k_1\, e_\phi(t) - k_2\, z(t) \cos\big(e_\phi(t)\big). \tag{7.22}$$

The state equation of the deviation is obtained by derivation of (7.19):

$$\dot{z}(t) = \mp \frac{d}{dt}\|e_p(t)\|$$
$$= \frac{\mp 1}{\|e_p(t)\|} e_p^T(t)\,\dot{e}_p(t)$$
$$= \frac{1}{z(t)}\Big(\underbrace{e_p^T(t)\,\dot{\tilde{p}}(t)}_{=\,0} - e_p^T(t)\,\dot{p}(t)\Big). \tag{7.23}$$

The first scalar product in the bracket is equal to zero because the movement of the projected position $\tilde{p}(t)$ is always directed along the path. Due to the orthogonal projection (7.5), the error $e_p(t)$ is orthogonal to the path and, therefore, also orthogonal to $\dot{\tilde{p}}(t)$. Application of (7.4) to (7.23) yields

$$\dot{z}(t) = -\frac{1}{z(t)}\,e_p^T(t)\,e\big(\phi(t)\big)\,v(t).$$

Analogously to (7.20), the relation

$$e_p^T(t)\,e\big(\phi(t)\big) = -z(t)\sin\big(e_\phi(t)\big)$$

is determined using (7.13) and (7.15). The derivative of $z(t)$ is then finally determined by

$$\dot{z}(t) = \sin\big(e_\phi(t)\big)\,v(t). \tag{7.24}$$

The combination of (7.22) and (7.24) yields the error model (7.21). ∎

The error dynamics of the robot (7.4) together with the lateral controller (7.12) is exactly described by the developed second-order error model (7.21) even though the robot has a higher order. This is due to the orthogonal projection which renders the elements of $e_p(t)$ linearly dependent on each other. Thus, the deviation of the robot from the path can be described by the scalar variable (7.19) as shown in (7.21). The next section will discuss how to find appropriate controller parameters k_1 and k_2 so that the error model (7.21) settles in the origin of its state space.

Stability and convergence

The stability of the error model Σ_e will be analysed with Lyapunov's Indirect Method [66]. The equilibria of Σ_e are

$$\bar{e}_\phi = n\,\pi, \quad n \in \mathbb{Z}, \tag{7.25}$$

which results from the first state equation of Σ_e with $\mathbb{Z}$ denoting the set of integers, and

$$\bar{z} = -\frac{k_1}{k_2}\frac{\bar{e}_\phi}{\cos(\bar{e}_\phi)}, \tag{7.26}$$

which is obtained from the second state equation. For the sake of simplicity, it is assumed that the commanded velocity input is restricted to constant values in the following:

$$u(t) \equiv \bar{u} > 0, \quad t \geq 0.$$

In doing so, the dynamical behaviour of Σ_e around each equilibrium is characterised by the Jacobian

$$A = \begin{pmatrix} 0 & \cos(\bar{e}_\phi)\,\bar{u} \\ -k_2\cos(\bar{e}_\phi) & -k_1 \end{pmatrix}$$

of (7.21) with the characteristic polynomial

$$p(\lambda) = \det(\lambda I - A) = \lambda^2 + k_1\lambda + k_2\,\bar{u}\,\underbrace{\cos^2(\bar{e}_\phi)}_{= 1, \text{ cf. (7.25)}},$$

where I denotes the identity matrix. Application of Routh–Hurwitz stability criterion reveals that each individual equilibrium point is asymptotically stable if and only if

$$k_1 > 0, \quad k_2 > 0 \tag{7.27}$$

hold true. Unfortunately, the error model Σ_e has an infinite number of equilibrium points and, hence, convergence to the desired equilibrium in the origin

$$\bar{z} = 0 \tag{7.28}$$
$$\bar{e}_\phi = 0 \tag{7.29}$$

can only be guaranteed if the robot is placed sufficiently close to the path initially. However, for controller parameters k_1 and k_2 in the same order of magnitude, the undesired equilibria (7.26) are at least multiple metres away from the path and, hence, did not pose any problems in the experiments.

Theorem 7.1 (Path tracking). *The lateral path controller (7.12) solves Problem 7.1 for $u(t) \equiv \bar{u} > 0, (t \geq 0)$ if and only if the controller parameters k_1 and k_2 are positive and the initial errors $z(0)$ and $e_\phi(0)$ are sufficiently small.*

Proof. First, observe that a robot that is initially located on the path with correct orientation

$$p(0) = \tilde{p}(0)$$
$$\phi(0) = \tilde{\phi}(0)$$

does not excite the error model (7.21) and, hence, the tracking error remains equal to zero for any $u(t), (t \geq 0)$. The control input (7.12) then reads as $\omega(t) = \dot{\tilde{\phi}}(t)$ and due to the second state equation of (7.4), it follows that

$$\dot{\phi}(t) = \dot{\tilde{\phi}}(t) \tag{7.30}$$

127

holds true. The result (7.30) show that the robot stays correctly oriented. Thus, the first state equation of (7.4) becomes

$$\dot{p}(t) = e\big(\tilde{\phi}(t)\big)\,\bar{u}, \tag{7.31}$$

which shows that the movement direction of the robot is always directed along the path and, hence, the robot stays on the path as soon as it arrives.

For nonzero initial errors $z(0)$ and $e_\phi(0)$ that are in a neighbourhood of the origin, the error model converges to the equilibrium (7.28), (7.29) if (7.27) holds true and, hence, (7.6) and (7.7) hold asymptotically. ∎

Even though the error model Σ_e becomes stationary in the equilibrium point (7.28) and (7.29), the path-controlled robot Σ will not approach a steady state as shown by (7.31) due to the high-level control input $u(t)$ which is exactly the desired behaviour. Since the orientation $\phi(t)$ is prescribed by the path, the path-controlled robot is described by the state equation (7.31) asymptotically, i. e. the lateral controller (7.12) also reduces the model complexity.

7.2.3 Construction of paths

The analysis provided in the last section is based on the proposed lateral controller (7.12) which explicitly requires the signals of the orthogonal projection $\tilde{p}(t)$ and $\tilde{\phi}(t)$ for implementation. Unfortunately, the calculation of the projection is not always possible in explicit form and highly depends on the type of path. Thus, the path will be assumed to be a composition of straight and circular segments in the following since it is possible to obtain an explicit form of the projection for these shapes. This is a mild restriction because usual paths can be constructed as a combination of straight and circular segments [110]. The segments are characterised as follows.

Straight segments. A straight path is characterised by the position vector p_0 and the angle ϑ. The projected orientation is trivially given by

$$\tilde{\phi}(t) \equiv \vartheta, \quad \dot{\tilde{\phi}}(t) \equiv 0 \tag{7.32}$$

since a straight path has no curvature. By introduction of the unit direction vector

$$r := \begin{pmatrix} \cos\vartheta \\ \sin\vartheta \end{pmatrix},$$

the projected position is given by

$$\tilde{p}(t) = p_0 + r\,s(t) \tag{7.33}$$

with $s(t)$ denoting the distance travelled on the segment. Insertion of (7.33) into the property (7.5) of the orthogonal projection yields

$$s(t) = (p(t) - p_0)^{\mathrm{T}} r,$$

which in turn results in

$$\tilde{p}(t) = p_0 + r(p(t) - p_0)^{\mathrm{T}} r \tag{7.34}$$

after insertion into (7.33). With the projection (7.32) and (7.34), the lateral controller (7.12) can be applied for straight paths.

Circular segments. A circular path is characterised by the centre point p_{m} and the radius R. The orthogonally projected position is obtained by radial scaling regarding the centre p_{m} which yields

$$\tilde{p}(t) = p_{\mathrm{m}} + R \frac{p(t) - p_{\mathrm{m}}}{\|p(t) - p_{\mathrm{m}}\|}. \tag{7.35}$$

The projected orientation angle is obtained by

$$\tilde{\phi}(t) = \arctan\left(\frac{y(t) - y_{\mathrm{m}}}{x(t) - x_{\mathrm{m}}}\right) \pm \frac{\pi}{2}, \tag{7.36}$$

which is the phase angle of the robot in polar coordinates regarding the centre p_{m} shifted by a quarter turn. The sign of $\pi/2$ determines the turning direction, i.e. $+\pi/2$ and $-\pi/2$ generate a counter-clockwise and clockwise turn, respectively. The derivative of $\tilde{\phi}(t)$ is

$$\begin{aligned}
\dot{\tilde{\phi}}(t) &= \left(1 + \left(\frac{y(t) - y_{\mathrm{m}}}{x(t) - x_{\mathrm{m}}}\right)\right)^{-1} \frac{\mathrm{d}}{\mathrm{d}t}\left(\frac{y(t) - y_{\mathrm{m}}}{x(t) - x_{\mathrm{m}}}\right) \\
&= \frac{\dot{y}(t)(x(t) - x_{\mathrm{m}}) - \dot{x}(t)(y(t) - y_{\mathrm{m}})}{(x(t) - x_{\mathrm{m}})^2 + (y(t) - y_{\mathrm{m}})^2} \\
&= -u(t) \frac{(p(t) - p_{\mathrm{m}})^{\mathrm{T}} n(\phi(t))}{\|p(t) - p_{\mathrm{m}}\|^2} \tag{7.37}
\end{aligned}$$

using the first state equation of the robot model (7.4). With the projection (7.35)–(7.37), the lateral controller (7.12) can be applied for circular paths.

Combination of straight and circular path segments

This section will discuss how to combine straight and circular segments to generate a path which can be tracked by a robot exactly. As shown by Theorem 7.1, a robot that converged to a segment will stay there. Thus, the transitions between adjacent segments have to be designed so that the error model (7.21) is not excited by switching the projection from one segment to the other. The following theorem provides a condition for that property.

Theorem 7.2 (Exact path tracking). *Let t_- and t_+ denote the points in time right before and after the transition between two adjacent segments of the path, respectively and suppose that the robot already converged to the path. If and only if the path is designed so that the projection satisfies*

$$\tilde{p}(t_+) = \tilde{p}(t_-), \tag{7.38}$$
$$\tilde{\phi}(t_+) = \tilde{\phi}(t_-), \tag{7.39}$$

the robot (7.4) steered by the lateral controller (7.12) will track the path exactly during and after the transition.

Proof. If the conditions (7.38) and (7.39) are violated, the position $p(t)$ or the orientation $\phi(t)$ has to jump. This is technically infeasible since the control inputs $u(t)$ or $\omega(t)$ would have to include a Dirac impulse at t_+ which proves the necessity of (7.38).

Sufficiency follows from the fact that the error model Σ_e is not excited if the condition (7.38) is satisfied since continuity of $\tilde{p}(t)$ and $\tilde{\phi}(t)$ implies continuity of $z(t)$ and $e_\phi(t)$ due to (7.8), (7.9) and (7.19). Thus, $\dot{z}(t_+) = 0$ and $\dot{e}_\phi(t_+) = 0$ hold true, i.e. the state of Σ_e stays continuously differentiable which proves the sufficiency of (7.38). ∎

Note that path tracking is achieved even though the derivative $\dot{\tilde{\phi}}(t)$ jumps during the transition from straight to circular paths or vice versa. This is due to the kinematics (7.4) which allows for discontinuities in the rotational velocity $\omega(t)$. However, real mobile systems will have some dynamics regarding the change of rotation which has been addressed in [131]. If the steering dynamics impair the path tracking performance, a practical solution is to initiate the steering a little bit beforehand to compensate for the lag.

The problem of combining different road or rail segments in the construction of roads or tracks is usually solved by application of track transition curves that gradually align the curvature of straight and circular segments so that the curvature is continuous. This can be done with clothoidal splines for example [43].

Example 7.1 (Construction of a path). An example path consisting of seven segments is depicted in Fig. 7.6. In order to construct the path, i.e. to find the parameters p_{0i}, ϑ_i, p_{mi} and R_i, $(i = 1, \ldots, 7)$, a set of equations can be obtained by (7.38) and (7.39) as follows.

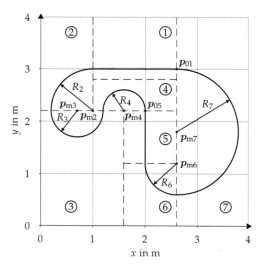

Fig. 7.6: Construction of the example path

The example path has two straight and five circular segments which are connected via the set of equations

$$x_{01} = x_{m7} \qquad\qquad y_{01} = R_2 + y_{m2}$$
$$x_{m2} = x_{m3} + R_2 - R_3 \qquad y_{m2} = y_{m3}$$
$$x_{m3} = x_{m4} - R_3 - R4 \qquad y_{m3} = y_{m4}$$
$$x_{m4} = x_{05} - R_4 \qquad\qquad y_{m4} = y_{05}$$
$$x_{05} = x_{m6} - R_6 \qquad\qquad y_{m6} = y_{m7} + R_6 - R_7$$
$$x_{m6} = x_{m7} \qquad\qquad y_{m7} = y_{01} - R_7.$$

There are 12 equation with with 19 unknown parameters. Thus, the system of equations is underdetermined and has an infinite number of solutions. To reduce the degrees of freedom, the parameters x_{01}, y_{01} and R_i, ($i \in \{2,3,4,6,7\}$) are fixed. The above set of equations can then be cast in the new system

$$
\begin{pmatrix}
0 & 1 & 0 & 0 & 0 & 0 & 0 & 0 & 0 & 0 & 0 & 0 \\
0 & -1 & 0 & 1 & 0 & 0 & 0 & 0 & 0 & 0 & 0 & 0 \\
1 & 0 & -1 & 0 & 0 & 0 & 0 & 0 & 0 & 0 & 0 & 0 \\
0 & 0 & -1 & 0 & 1 & 0 & 0 & 0 & 0 & 0 & 0 & 0 \\
0 & 0 & 0 & -1 & 0 & 1 & 0 & 0 & 0 & 0 & 0 & 0 \\
0 & 0 & 0 & 0 & 0 & -1 & 0 & 1 & 0 & 0 & 0 & 0 \\
0 & 0 & 0 & 0 & -1 & 0 & 1 & 0 & 0 & 0 & 0 & 0 \\
0 & 0 & 0 & 0 & 0 & 0 & -1 & 0 & 1 & 0 & 0 & 0 \\
0 & 0 & 0 & 0 & 0 & 0 & 0 & 0 & -1 & 0 & 1 & 0 \\
0 & 0 & 0 & 0 & 0 & 0 & 0 & 0 & 0 & 1 & 0 & -1 \\
0 & 0 & 0 & 0 & 0 & 0 & 0 & 0 & 0 & 0 & 0 & 1 \\
0 & 0 & 0 & 0 & 0 & 0 & 0 & 0 & 0 & 0 & 1 & 0
\end{pmatrix}
\begin{pmatrix}
x_{m2} \\ y_{m2} \\ x_{m3} \\ y_{m3} \\ x_{m4} \\ y_{m4} \\ x_{05} \\ y_{05} \\ x_{m6} \\ y_{m6} \\ x_{m7} \\ y_{m7}
\end{pmatrix}
=
\begin{pmatrix}
y_{01} - R_2 \\ 0 \\ R_2 - R_3 \\ R_3 + R_4 \\ 0 \\ 0 \\ R_4 \\ R_6 \\ 0 \\ R_6 - R_7 \\ y_{01} - R_7 \\ x_{01}
\end{pmatrix},
$$

which can be solved straightforwardly by inversion of the (12×12)-matrix to obtain the remaining parameters that are given in Table 7.1. □

Table 7.1: Parameters of the path

Segment	①	②	③	④
	$p_{01} = \begin{pmatrix} 2.6\,\text{m} \\ 3\,\text{m} \end{pmatrix}$	$p_{m2} = \begin{pmatrix} 1\,\text{m} \\ 2.2\,\text{m} \end{pmatrix}$	$p_{m3} = \begin{pmatrix} 0.7\,\text{m} \\ 2.2\,\text{m} \end{pmatrix}$	$p_{m4} = \begin{pmatrix} 1.6\,\text{m} \\ 2.2\,\text{m} \end{pmatrix}$
	$\vartheta_1 = \pi$	$R_2 = 0.8\,\text{m}$	$R_3 = 0.5\,\text{m}$	$R_4 = 0.4\,\text{m}$

$\cdots$	⑤	⑥	⑦
	$p_{05} = \begin{pmatrix} 2\,\text{m} \\ 2.2\,\text{m} \end{pmatrix}$	$p_{m6} = \begin{pmatrix} 2.6\,\text{m} \\ 1.2\,\text{m} \end{pmatrix}$	$p_{m7} = \begin{pmatrix} 2.6\,\text{m} \\ 1.8\,\text{m} \end{pmatrix}$
	$\vartheta_5 = 1.5\,\pi$	$R_6 = 0.6\,\text{m}$	$R_7 = 1.2\,\text{m}$

7.2.4 Design of a velocity controller

The path-controlled robot with the structure in Fig. 7.4 only fixes the rotational velocity $\omega(t)$. The commanded velocity $u(t)$ is free to choose and should be assumed by the robot asymptotically:

$$\lim_{t \to \infty} |u(t) - v(t)| = 0. \tag{7.40}$$

Due to uncertainties in the model, measurement noise and manufacturing tolerances of the robots, a velocity controller will be developed in this section to achieve (7.40). To this end, consider the response of the path-controlled robot to the input $u(t) = \bar{u}\,\sigma(t - 0.5\,\text{s})$, $(t \geq 0)$ for different values of $\bar{u}$ shown in Fig. 7.7.

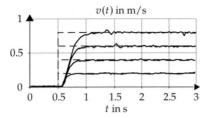

Fig. 7.7: Measured step response (solid) for different reference steps $\bar{u}$ (dashed) of the path-controlled robot in Fig. 7.4

Note that the response of the path-controlled robot without a velocity controller already satisfies (7.40) quite well. However, the behaviour shown in Fig. 7.7 is not robust in the presence of disturbances. Furthermore, the response of the robot is very steep and, thus, the robot is prone to wheel slipping without a controller that is able to slow down the response.

In a first approximation, the dynamical behaviour of a path-controlled robot can be described by the first-order dynamics

$$\Sigma:\ V(s) = G(s)\,U(s) \tag{7.41}$$

with the transfer function

$$G(s) = \frac{1}{\tau s + 1} \tag{7.42}$$

with the time lag $\tau = 150\,\text{ms}$ identified from the measurement in Fig. 7.7. In order to achieve (7.40), the output feedback

$$\Sigma_R:\ U(s) = K(s)\left(V_0(s) - V(s)\right)$$

with $V_0(s)$ •—○ $v_0(t)$ denoting the reference velocity is applied with the PI-type controller transfer function

$$K(s) = k_P + \frac{k_I}{s}. \tag{7.43}$$

In addition to asymptotic stability, the parameters k_P and k_I should be chosen so that the velocity and path-controlled robot is externally positive as will be elaborated in the following using the results of Chapter 3.

The closed-loop transfer function (cf. (3.10))

$$\bar{G}(s) = \frac{G(s)\,K(s)}{1 + G(s)\,K(s)} = \frac{k_P\,s + k_I}{\tau s^2 + (1 + k_P)\,s + k_I} \tag{7.44}$$

133

has a transmission zero

$$s_0 = -\frac{k_I}{k_P}$$

and two poles

$$s_{1/2} = -\frac{1 + k_P}{2\tau} \pm \sqrt{\left(\frac{1 + k_P}{2\tau}\right)^2 - \frac{k_I}{\tau}}. \tag{7.45}$$

The closed-loop system is asymptotically stable and externally positive if and only if the poles are real and negative and if $\bar{G}(s)$ has a positive decomposition of the form (3.7). That is, the zero has to be located to the left of the dominant pole:

$$s_0 \overset{!}{<} s_1. \tag{7.46}$$

These objectives are satisfied if and only if the parameters are chosen according to

$$0 < k_I < \left(\frac{\sqrt{\alpha - 1} - \sqrt{\alpha}}{\sqrt{\tau}\left(1 - \sqrt{4\alpha}\left(\sqrt{\alpha} - \sqrt{\alpha - 1}\right)\right)}\right)^2 \tag{7.47}$$

$$k_P = \sqrt{4\tau\alpha k_I} - 1 \tag{7.48}$$

for some $\alpha \geq 1$. The conditions (7.47) and (7.48) are obtained straightforwardly by imposing nonnegativity on the term under the square root in (7.45) and consideration of (7.46). The design procedure is summarised in Algorithm 7.1.

Algorithm 7.1 (Externally positive velocity controller).

Input: Time lag τ of a path-controlled robot.

1. Choose an $\alpha \geq 1$.
2. Choose the parameter k_I that satisfies (7.47).
3. Determine the parameter k_P with (7.48).

Output: Asymptotically stable and externally positive velocity and path-controlled robot with the transfer function (7.44).

Algorithm 7.1 offers some freedom to shape the behaviour beyond the stability and external positivity. For $\alpha = 1$, both poles have the same value. For increasing values of α, the poles split into two distinctive real poles. The interpretation of the upper bound of k_I in (7.47) is similar. For a large k_I, the transmission zero s_0 approaches the dominant eigenvalue s_1 and a small nonnegative choice of k_I increases the gap between s_0 and s_1.

7.2.5 Experimental evaluation of the path controller

The following experiments are carried out with the PI-type velocity controller (7.43) as the longitudinal controller combined with the lateral controller (7.12) with the parameters $k_1 = 5$ and $k_2 = 8$. The resulting control structure is shown in Fig. 7.8.

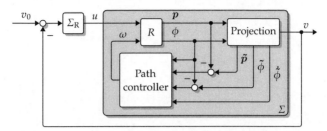

Fig. 7.8: Cascaded control loop of a velocity and path-controlled robot

Figure 7.9 shows the example path (solid line) and the trajectories (dashed lines) of the controlled robot on the driving surface starting from different initial conditions (circles). The longitudinal velocity controller fixes the velocity of the robot to $v_0(t) \equiv \bar{v} = 0.3 \, \text{m/s}$. Clearly, the robot approaches the path from any initial condition and stays on the path once it arrived which verifies Theorem 7.1. Furthermore, the measurements show that the robot approaches the path quickly and without oscillating along the path for an appropriate choice of controller parameters k_1 and k_2.

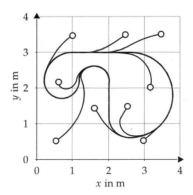

Fig. 7.9: Measured trajectories (solid) of a path-controlled robot starting from different initial states (circles o)

The convergence of the robot to the path has been proved for constant reference velocities in Theorem 7.1. In the next measurement, the performance of the proposed lateral controller (7.12) will be evaluated with a time varying velocity profile. The velocity profile is shown in Fig. 7.10. The dashed line represents the reference velocity $v_0(t)$ and the solid line is the measured velocity $v(t)$ that is controlled by the longitudinal controller.

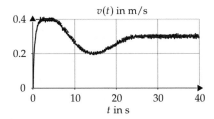

Fig. 7.10: Measured velocity (solid) and reference (dashed)

Figure 7.11 (a) shows the course of the coordinates $x(t)$ and $y(t)$ (solid lines) together with the corresponding projection (dashed lines) as well as the deviation $z(t)$ in the bottom graph for the initial condition

$$x_0 = 2.5\,\text{m}, \quad y_0 = 3.5\,\text{m}, \quad \phi_0 = 1.5\,\pi. \tag{7.49}$$

It can be seen that once the robot reaches the path in the first few seconds, it stays there through all transitions. Note that the error $z(t)$ shows a small deviation from the zero line (especially after the transition from segment ③ to ④) due to model uncertainties. It can be observed that the time varying velocity profile due to the longitudinal controller does not affect the performance of the lateral controller since it has no impact on the course of the error $z(t)$.

The corresponding orientation angle $\phi(t)$ (solid line) together with the projection (dashed line) are given in Fig. 7.11 (b) with the orientation error $e_\phi(t)$ illustrated in the bottom graph. Again, as soon as the robot approaches the path, it tracks the path correctly through the transitions. The error $e_\phi(t)$ shows small spikes after each transition which is due to the neglected dynamics of the rotation of the robot. However, the lateral controller achieves satisfactory performance so that Theorem 7.2 is considered verified. Furthermore, the conducted experiments show that the proposed lateral controller together with the explicit projection approach are real-time capable.

7.2.6 Summary of path control

This section considered path tracking of mobile systems and presented a low-level lateral controller based on an orthogonal projection that complements high-level longitudinal

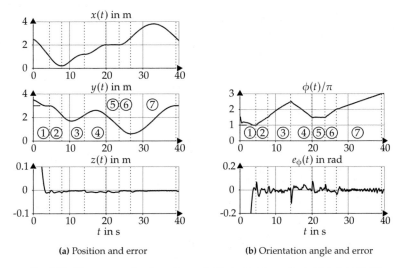

(a) Position and error   **(b)** Orientation angle and error

Fig. 7.11: Measurements of a path-controlled robot from the initial state (7.49)

controllers. It has been shown that the proposed controller stabilises the error model for appropriately chosen controller parameters. Furthermore, it has been discussed how to construct paths with straight and circular segments so that the lateral controller is able to track the path exactly through the transitions. To this end, explicit formulas to determine the orthogonal projection for these classes of path segments are given which simplify the implementation of the controller. In the laboratory experiments, the lateral controller achieved satisfactory tracking performance through all transitions between the path segments.

7.3   Vehicle platooning

7.3.1   Aim of the experiments

As shown in different studies with real vehicles (e. g. [47] and [98]), commercially available adaptive cruise controllers do not render the controlled vehicles externally positive. Instead, $\mathcal{L}_2$ string stability is usually imposed in the design of (cooperative) adaptive cruise controllers (ACC/CACC) in various references as [64, 99, 103, 114, 115, 146]. However, the results in [84] and the Example 4.1 in Chapter 4 show that $\mathcal{L}_2$ string stability is not sufficient for collision avoidance.

The experiments in the following will verify that a platoon of $\mathcal{L}_2$ string stable vehicles will end up in a collision if they are not externally positive to evaluate the theoretical results. During the experiment, the reference velocity $v_0(t)$ changes according to

$$v_0(t) = \begin{cases} 0.1\,\text{m/s}, & 0\,\text{s} \leq t < 5\,\text{s} \\ 0.7\,\text{m/s}, & 5\,\text{s} \leq t < 20\,\text{s} \\ 0.1\,\text{m/s}, & 20\,\text{s} \leq t < 35\,\text{s} \\ 0.5\,\text{m/s}, & 35\,\text{s} \leq t \end{cases} \tag{7.50}$$

in order to study the behaviour of a platoon with $N = 10$ robots in two cases. First, adaptive cruise controllers are applied that render the controlled robots $\mathcal{L}_2$ string stable but *not* externally positive. In the second case, the adaptive cruise controllers render each controlled robot externally positive.

7.3.2 Combined lateral and longitudinal control

The experiments will use the lateral controller that has been developed in Section 7.2 with which the robots are able to track a circular path. The leader robot $i = 1$ uses the velocity controller proposed in Section 7.2.4 as the longitudinal controller to follow the reference (7.50). The follower robots $i = 2, 3, \ldots, N$ use an adaptive cruise controller to control the inter-robot distance in the following experiments.

The circular path is characterised by the radius $R = 1.6\,\text{m}$ as well as the middle point

$$p_{\text{m}} = \begin{pmatrix} x_{\text{m}} \\ y_{\text{m}} \end{pmatrix} = \begin{pmatrix} 2\,\text{m} \\ 2\,\text{m} \end{pmatrix}$$

and allows the platoon to travel arbitrarily long. For a circular path, the projection is given explicitly by (7.35)–(7.37) in Section 7.2.3. Since the velocity $v_i(t)$ of a robot is not directly measurable, it is obtained as follows. The arc length

$$s_i(t) = R\,\tilde{\phi}_i(t),$$

which represents the travelled distance of the projected position $\tilde{p}_i(t)$ on the circular path, can be either differentiated numerically or the velocity can be calculated by

$$v_i(t) = R\,\dot{\tilde{\phi}}_i(t)$$

using (7.37).

7.3.3 Platoon with $\mathcal{L}_2$ string stable vehicles

The path-controlled robot (7.41) with the transfer function (7.42) should be rendered $\mathcal{L}_2$ string stable by the choice of an appropriate feedback. For that purpose, consider the distance dynamics (cf. (4.4))

$$D_i(s) = s^{-1}\left(V_{i-1}(s) - V_i(s)\right). \tag{7.51}$$

The local control law

$$U_i(s) = K(s)\left(\beta V_i(s) - D_i(s)\right) \tag{7.52}$$

with $K(s)$ denoting the controller transfer function adjusts the time-headway error (cf. (R2) on page 41). The combination of (7.42), (7.51) and (7.52) yields the closed-loop transfer function

$$\bar{G}(s) = \frac{G(s)\,K(s)}{G(s)\,K(s)\,(\beta s + 1) - s}. \tag{7.53}$$

The distance transfer function

$$G_d(s) = \frac{G(s)\,K(s)\,\beta - 1}{G(s)\,K(s)\,(\beta s + 1) - s} \tag{7.54}$$

results from the relation (4.5). The transfer functions (7.53) and (7.54) characterise the closed-loop dynamics $\bar{\Sigma}_i$ (cf. (4.2)) completely in terms of the controller $K(s)$.

The controller transfer function is chosen as the PI-type controller

$$K(s) = k_P + \frac{k_I}{s}. \tag{7.55}$$

With this choice, it is straightforward to verify $\bar{G}(0) = 1$, $G_d(0) = \beta$, i.e. the design objectives (D2) and (D3) (cf. page 50) are satisfied structurally.

Table 7.2: $\mathcal{L}_2$ string stable configuration

Parameter	k_P	k_I	β
Value	−0.05	−5.6	0.6 s

The impulse response and the magnitude plot shown in Fig. 7.12 result from the parameters given in Table 7.2. The impulse response in Fig. 7.12 (a) is negative in a time interval around $t = 2$ s and vanishes asymptotically. Hence, the controlled robot is asymptotically stable but it is *not* externally positive. However, as can be seen in the magnitude plot in Fig. 7.12 (b), the condition on $\mathcal{L}_2$ string stability provided by Lemma 4.1 is satisfied since the magnitude is bounded by $\bar{G}(0) = 1$.

In summary, the controller configuration in Table 7.2 to parametrise the feedback (7.52) with the PI-type controller (7.55) satisfies the design objectives (D1)–(D3) but instead of (D4), the controlled robot is rendered $\mathcal{L}_2$ string stable.

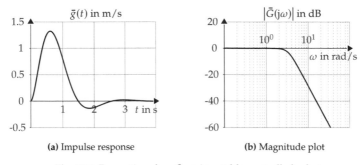

(a) Impulse response   (b) Magnitude plot

Fig. 7.12: Properties of an $\mathcal{L}_2$ string stable controlled robot

Measurements

The measured velocity and distance responses of the $\mathcal{L}_2$ string stable platoon of $N = 10$ robots to the reference input (7.50) of the leader is shown in Fig. 7.13. The dashed line represents the reference $v_0(t)$ or $\beta\,v_0(t)$ in the top or bottom graph, respectively, and the solid lines represent the velocity (top) or inter-robot distance (bottom) of each robot in the platoon.

The follower robots clearly show overshooting responses in both the velocity $v_i(t)$ and the distance $d_i(t)$. Furthermore, the amplitudes of the overshooting responses build up along the string. It can be seen that the velocity and the distance of each robot assume their references asymptotically from which it can be concluded that the requirements (R1) and (R2) (cf. page 41) are satisfied. However, at the braking instance at $t = 20\,\mathrm{s}$, the overshooting responses cause a collision at around $t = 27\,\mathrm{s}$ where the inter-robot distance $d_i(t)$ reaches zero, i. e. the minimum distance $d_0 = 0.18\,\mathrm{m}$ is not complied with at this point in time (cf. (4.1)). The same holds true for the velocity which reaches zero at the same point in time. Adding an additional robot to the platoon would result in the signals to become negative. Thus, a platoon with $\mathcal{L}_2$ string stable vehicles does not satisfy the requirements (R3) and (R4).

A similar behaviour has been observed in the experiments presented in [98] where real vehicles with commercially available ACC systems have been studied. When the leader vehicle brakes very hard, the followers undershoot the desired inter-vehicle distance and come dangerously close to their predecessor. In some cases, the ACC system even shuts down and leaves the situation to the driver.

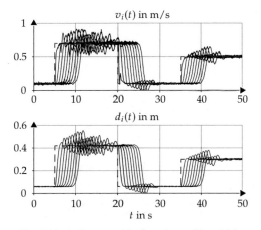

Fig. 7.13: A platoon with $\mathcal{L}_2$ string stable vehicles

7.3.4 Platoon with externally positive vehicles

A path-controlled robot with the transfer function (7.42) can be described by the first-order system

$$\Sigma_i : \ \dot{v}_i(t) = -\frac{1}{\tau}\,v_i(t) + \frac{1}{\tau}\,u_i(t), \quad v_i(0) = v_{i0} \tag{7.56}$$

with the time constant $\tau = 150\,\text{ms}$. The model (7.56) has the required linear form (4.25) and satisfies Assumption 4.1 as well as Assumption 4.2. Thus, Algorithm 4.2 can be applied to find the gain k of the feedback

$$u_i(t) = -\boldsymbol{k}^{\mathrm{T}} \begin{pmatrix} v_i(t) \\ d_i(t) \\ z_i(t) \end{pmatrix} \tag{7.57}$$

with the controller state $z_i(t)$ defined by (4.31). By application of Ackermann's formula (4.39), the feedback gain is explicitly given by

$$\boldsymbol{k} = \begin{pmatrix} -\tau\,(\bar{\lambda}_1 + \bar{\lambda}_2 + \bar{\lambda}_3) - 1 \\ -\tau\,(\beta\,\bar{\lambda}_1\bar{\lambda}_2\bar{\lambda}_3 + \bar{\lambda}_1\bar{\lambda}_2 + \bar{\lambda}_2\bar{\lambda}_3 + \bar{\lambda}_1\bar{\lambda}_3) \\ -\tau\,\bar{\lambda}_1\bar{\lambda}_2\bar{\lambda}_3 \end{pmatrix}.$$

The closed-loop eigenvalues $\bar{\lambda}_i$, ($i \in \{1,2,3\}$) are obtained with Algorithm 4.1 which is executed with $\beta = 0.6\,\mathrm{s}$ and $n = 1$ and results in the closed-loop eigenvalues

$$\bar{\lambda}_1 \in \left(-\frac{2}{\beta}, -\frac{1}{\beta}\right) = \left(-3.33\,\mathrm{s}^{-1}, -1.67\,\mathrm{s}^{-1}\right)$$

$$\bar{\lambda}_2 = -\frac{\bar{\lambda}_1}{\beta\bar{\lambda}_1 + 1}$$

$$\bar{\lambda}_3 = \bar{\mu}.$$

By choosing $\bar{\lambda}_1 = -2.5\,\mathrm{s}^{-1}$, the second eigenvalue is set to $\bar{\lambda}_2 = -5\,\mathrm{s}^{-1}$. The closed-loop transmission zero is chosen according to $\bar{\mu} = -6\,\mathrm{s}^{-1}$ which determines the last closed-loop eigenvalue. This set of eigenvalues results in the state feedback gain

$$k^{\mathrm{T}} = \begin{pmatrix} k_{\mathrm{v}} & k_{\mathrm{d}} & k_{\mathrm{z}} \end{pmatrix} = \begin{pmatrix} 1.025 & -1.875 & 11.250 \end{pmatrix}.$$

Note that the control law (7.57) can be implemented as an output feedback using a PI-type controller since the path-controlled robot Σ_i is represented by the first-order model (7.56) and, hence, its state is composed solely of the velocity $v_i(t)$. Thus, the feedback unit is given by

$$F_i : \begin{cases} \dot{z}_i(t) = \begin{pmatrix} \beta & -1 \end{pmatrix} \begin{pmatrix} v_i(t) \\ d_i(t) \end{pmatrix}, & z_i(0) = z_{i0} \\[2ex] u_i(t) = -k_{\mathrm{z}} z_i(t) - \begin{pmatrix} k_{\mathrm{v}} & k_{\mathrm{d}} \end{pmatrix} \begin{pmatrix} v_i(t) \\ d_i(t) \end{pmatrix}. \end{cases}$$

Measurements

In the following, two measurements are shown. First, the impulse response of a controlled robot is measured to show that the robot is indeed rendered externally positive by Algorithm 4.2 and, second, a platoon of $N = 10$ robots with the changing reference (7.50) of the leader is examined. Since the impulse response is hardly measurable directly, it will be obtained with a trick. The unforced (free) motion of a controlled robot is determined by

$$v_{i,\text{free}}(t) = \bar{c}^{\mathrm{T}} e^{\bar{A}t} \bar{x}_{i0} \tag{7.58}$$

with the elements of the closed-loop model (4.34) and the initial state $\bar{x}_{i0}^{\mathrm{T}} = \begin{pmatrix} v_{i0} & d_{i0} & z_{i0} \end{pmatrix}$. The unforced motion (7.58) differs only in the last factor from the impulse response

$$\bar{g}(t) = \bar{c}^{\mathrm{T}} e^{\bar{A}t} e.$$

Thus, the impulse response $\bar{g}(t)$ of the controlled robot can be obtained by measuring its velocity $v_i(t)$ with $v_{i-1}(t) \equiv 0$ if the initial state is set according to

$$\begin{pmatrix} v_{i0} \\ d_{i0} \\ z_{i0} \end{pmatrix} = \begin{pmatrix} 0 \\ 1\,\mathrm{m} \\ 0 \end{pmatrix} = e.$$

with e defined in (4.32). The result is depicted in Fig. 7.14 which shows that the closed-loop impulse response is nonnegative for all t, i.e. the proposed design procedure satisfies the design goal in an experimental environment.

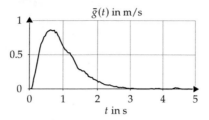

Fig. 7.14: Measured impulse response

Figure 7.15 shows the response of a platoon with externally positive robots to the reference (7.50). Again, the dashed line represents the reference $v_0(t)$ or $\beta\,v_0(t)$ in the top or the bottom graph, respectively, and the solid lines represent the measured velocity (top) or the inter-robot distance (bottom) of each robot in the platoon.

It can be seen that after each reference step, all robots travel with the correct velocity after all transients settled. The transients are locally monotonic due to the externally positive dynamics. Thus, neither the velocity $v_i(t)$ nor the inter-robot distance $d_i(t)$ show any overshoots during the transition to the new set-point of the reference $v_0(t)$. In particular, the braking manoeuvre at $t = 20\,$s does not cause a collision as in Fig. 7.13 with $\mathcal{L}_2$ string stable robots due to the monotonicity of the response.

The measurements show that the requirements (R1)–(R4) are satisfied by the platoon with the controlled robots and the feedback gain k designed by the method proposed in Chapter 4.

7.3.5 Discussion of the results

Both measurements satisfy requirements (R1) and (R2) as the velocity $v_i(t)$ and the inter-robot distance $d_i(t)$ approach the set-point asymptotically. The platoon with $\mathcal{L}_2$ string stable robots (Fig. 7.13) clearly shows significant hazardous behaviour due to the overshooting responses and as observed in the measurements, the requirements (R3) and (R4) are *not* satisfied.

In contrast, the platoon with externally positive robots (Fig. 7.15) performs smoothly. The set-point is reached monotonically without any overshoot due to the nonnegative impulse response. Thus, the requirements (R1)–(R4) are satisfied for any number N of robots in the platoon. The difference in the characteristics given in Figs. 7.12 and 7.14 is very small but the transient behaviour differs significantly as illustrated by the measurements in Figs. 7.13 and 7.15.

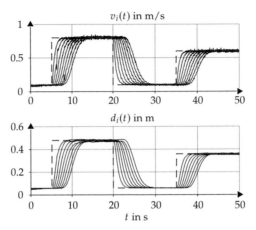

Fig. 7.15: A platoon with externally positive vehicles

The fact that a lot of studies prefer $\mathcal{L}_2$ string stability as a safety measure results from the circumstance that it is easier to render a controlled vehicle $\mathcal{L}_2$ string stable with Lemma 4.1 than to render it externally positive. Furthermore, the characteristics in Fig. 7.12, which result in a collision in the platoon when the leader is braking, is a narrow special case. A conservative approach to the design of $\mathcal{L}_2$ string stable closed-loop dynamics might end up in external positivity as can be observed in [64, 104, 146]. The controller structure and design method that is proposed in Chapter 4 solves this problem by providing a systematic way to render the controlled vehicles externally positive.

7.3.6 Extension: Merging of multiple platoons

The platooning controller that has been developed in Chapter 4 has been evaluated experimentally in the last section. Platooning is an important fundamental problem in the context of autonomous driving. However, there are situation that require the cooperation of multiple platoons, for example, to pass a lane reduction. To this end, this section will concern *merging* as an extension of platooning to extend the previous results.

Merging describes the problem of combining two adjacent platoons with each other immediately before a lane reduction to obtain a single platoon as illustrated in Fig. 7.16. In dense traffic, the vehicles should use the zipper method. That is, the vehicles pass the lane reduction alternately from the *main* (top) and the *merging* (bottom) lane. Advances in the development of autonomous vehicles (e. g. Google or Tesla cars) enable the vehicles to change the lane automatically if there is a sufficiently large gap between adjacent

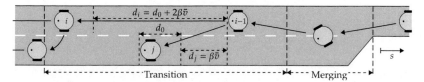

Fig. 7.16: Illustration of the merging procedure. Solid arrows indicate the information flow and distances are depicted with dashed arrows.

vehicles on the target lane. However, in dense traffic, the vehicles cannot rely on randomly occurring gaps between adjacent vehicles to steer in. Thus, it is necessary for the vehicles to cooperate with each other. To this end, a control concept based on different phases which correspond to a segmentation of the road is proposed and tested experimentally in the following with mobile robots. All robots are equipped with multiple controllers that handle the different segments which are characterised as follows:

1. **Platooning:** In this segment, the robots are organised in strings of arbitrary length. The adaptive cruise controllers adjust the individual velocity in order to maintain a prescribed inter-robot distance as discussed in Chapter 4 and Section 7.3.

2. **Transition:** Each time a robot j on the merging lane enters the transition region, it has to adapt its inter-robot distance to its new predecessor $i-1$ on the main lane. In the same time, the follower i of the common predecessor $i-1$ has to reduce its velocity to create a gap on the main lane. This process is executed by the merging controller and has to be finished before the merging robot j enters the merging region.

3. **Merging:** After the transition, it is safe for the merging robot j to change to the main lane. At this point, it is part of the platoon on the main lane and the adaptive cruise controller is reactivated.

Let l_e be the length of the transition region. The transition phase has to be completed in a time span not longer than $\Delta t = t_e - t_0 = l_e/\bar{v}$ so that the generation and the alignment to the emerging gap is finished before robot j enters the merging region. In order to accomplish the adjustment of the distances in finite time, a trajectory tracking controller will be applied.

Cooperative trajectory tracking

The goal of the merging controller is to adjust the inter-robot distance in finite time by application of a feedforward controller that is able to perform exact trajectory tracking [53]. Consider the robot j on the merging lane as the ego robot. Once robot j enters the transition region at $t = t_{j0}$, the feedforward controller has to work in cooperation with other robots due to the following observation. The distance dynamics

$$\dot{d}_j(t) = v_{i-1}(t) - v_j(t), \quad d_j(t_{j0}) = s_{i-1}(t_{j0}) - s_j(t_{j0}) - d_0 \tag{7.59}$$

is governed by the difference of the velocities of the ego robot j and its new predecessor $i-1$ on the main lane. Thus, given a desired trajectory $d_j^\star(t)$, the merging controller has to consider the velocity $v_{i-1}(t)$ of the predecessor to accomplish its task of tracking the distance trajectory

$$d_j(t) \overset{!}{=} d_j^\star(t), \quad t \in [t_{0j}, t_{ej}]$$

in a specific interval with t_{0j} and t_{ej} denoting the points in time in which the robot j entered the transition region and when the transition has to be finished in order to respect the length l_e of the transition region. To this end, the distance dynamics (7.59) is rearranged to obtain the local trajectory

$$v_j^\star(t) = v_{i-1}^{com}(t) - \dot{d}_j^\star(t)$$

of the velocity which has to be implemented by the the feedforward unit F_i so that

$$v_j(t) \overset{!}{=} v_j^\star(t), \quad t \in [t_{0j}, t_{ej}] \tag{7.60}$$

holds true. The *communicated* velocity $v_{i-1}^{com}(t)$ of the predecessor is obtained via digital communication and, hence, marked by the superset com to highlight the distinction from the physical signal that works on the robot via the sensor network.

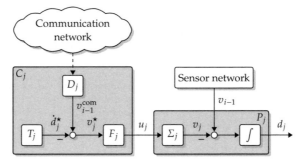

Fig. 7.17: Block diagram of the cooperative feedforward controller

The resulting control structure is illustrated in Fig. 7.17. The local controller C_j is complemented by the trajectory planning unit T_j that generates the desired trajectory $d_j^\star(t), (t \in [t_{0j}, t_{ej}])$. The feedforward unit F_j is synthesized by inversion-based or flatness-based control techniques in order to achieve (7.60). For the model Σ_j in (7.56), an inversion-based feedforward controller is given by

$$F_j : \ u_j(t) = \tau \, \dot{v}_j^{\star}(t) + v_j^{\star}(t).$$

While the robot j on the merging lane adjusts its distance to the new predecessor $i-1$, the robot i (cf. Fig. 7.16) also adjusts its inter-robot distance in the same time interval

$$d_i(t) \overset{!}{=} d_i^{\star}(t), \quad t \in [t_{0j}, t_{ej}]$$

to generate a gap simultaneously to the alignment of the merging robot j using the same control technique. The design of appropriate trajectories $d_i^{\star}(t)$ and $d_j^{\star}(t)$ will be discussed in the following.

Trajectory planning

The trajectories $d_i^{\star}(t)$ and $d_j^{\star}(t)$ are planned locally by the units T_i and T_j and have to be known before the transition phase begins at t_{0j}. The requirements

$$d_i^{\star}(t_{0j}) = s_{i-1}(t_{0j}) - s_i(t_{0j}) - d_0, \qquad d_j^{\star}(t_{0j}) = s_{i-1}(t_{0j}) - s_j(t_{0j}) - d_0, \quad (7.61)$$

$$d_i^{\star(k)}(t_{0j}) = v_{i-1}^{(k-1)}(t_{0j}) - v_i^{(k-1)}(t_{0j}), \qquad d_j^{\star(k)}(t_{0j}) = v_{i-1}^{(k-1)}(t_{0j}) - v_j^{(k-1)}(t_{0j}), \quad (7.62)$$

$$d_i^{\star}(t_{ej}) = \delta_{ei}, \qquad\qquad\qquad d_j^{\star}(t_{ej}) = \delta_{ej}, \quad (7.63)$$

$$d_i^{\star(k)}(t_{ej}) = 0, \qquad\qquad\qquad d_i^{\star(k)}(t_{ej}) = 0 \quad (7.64)$$

for $k = 1, 2, \dots, r$ are imposed on the trajectory with r denoting the relative degree of the model P_i. The first two requirements (7.61) and (7.62) ensure state consistency and follow from the fact that the inter-robot distance cannot jump. Requirement (7.63) determines the final distance and (7.64) makes sure that all velocities are equal again after the transition phase at $t = t_{ej}$. These requirements can be satisfied with the polynomials

$$d_i^{\star}(t) = \sum_{k=0}^{K} p_{ki} \, t^k, \qquad\qquad d_j^{\star}(t) = \sum_{k=0}^{K} p_{kj} \, t^k$$

of order $K = 2r + 1$.

Figure (7.18) shows an example of a typical trajectory for the merging robot j. Since the robots naturally congest before the lane reduction, the distance is decreasing initially. After the transition time t_{ej}, the merging controller should establish the final distance (7.63) whose value depends on the considered robot. The robot on the main lane that creates the gap (robot i in Fig. 7.16) has to double the distance, i. e. $\delta_{ei} = d_0 + 2\beta \bar{v}$ and the merging robot j adjusts the distance so that it equals the set-point $\delta_{ej} = \beta \bar{v}$ from the time-headway policy (cf. requirement (R2) on page 41).

In addition to the requirements (7.61)–(7.64), the trajectories have to be nonnegative in order to guarantee collision avoidance. For the model Σ_j given by (7.56), a condition on the choice of t_{ej} has been elaborated in [2] which ensures nonnegativity of the trajectory.

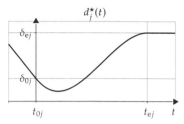

Fig. 7.18: Typical trajectory

Measurements

The effectiveness of the proposed methods is evaluated by an experiment with a platoon of five robots on the main lane and four robots on the merging lane. The measurements combine the path controller presented in Section 7.2 with the platooning controller of Chapter 4 and the cooperative tracking controller presented in this section. At first, the platooning controller that has been designed in Section 7.3 adjusts the velocity of the platoon to the reference velocity $\bar{v} = 0.45\,\mathrm{m/s}$ and, then, the cooperative tracking controllers perform the adjustments of the distances for the merging manoeuvre. Afterwards, the adaptive cruise controllers work in a single platoon. The minimum distance is given by $d_0 = 0.18\,\mathrm{m}$ for the considered robots.

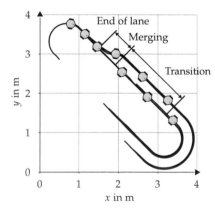

Fig. 7.19: Trajectories on the plane with robot positions at $t = 14\,\mathrm{s}$

Figure 7.19 shows the trajectories of all robots on the plane with their position at a fixed point in time. It can be seen that the distance between the robots on the main lane in the transition region is increased so that a gap is generated which is large enough for an additional robot from the merging lane to steer in. After the transition phase, the platooning controller is reactivated with a new predecessor and the position of the merging robot is projected on the main lane so that the path controller performs the lane change. Since the area of the driving surface is limited, the robots are guided along half circles before and after the transition region and the merging region in order to extend the usable path. The path is constructed with straight and circular elements as proposed in Section 7.2.3.

Figure 7.20 (a) shows the velocities and the inter-robot distances of all robots. The solid and the dashed lines represent robots on the main and the merging lane, respectively. The merging controllers decelerate the cooperating robots during the transition phase as depicted in the top graph. The bottom graph of Fig. 7.20 (a) shows the inter-robot distances of all robots. Each time a robot on the merging lane (depicted by dashed lines) enters the transition region, it finds a new predecessor on the main lane and the distance jumps to a new lower value. The reduced velocity then results in an increase of the inter-robot distance which generates the gaps on the main lane and aligns the merging vehicles to the gaps. Note that even though some distances fall below the minimum distance d_0 shortly, there are no collisions since the corresponding robots drive on different lanes as long as they are in the transition phase.

Figure. 7.20 (b) shows the projected positions of all robots. Again, the dashed lines represent robots on the merging lane. The transition region and the merging region begin at 3.42 m and 5.52 m, respectively. This leads to a transition length of $l_e = 2.1$ m and, thus, yields the maximum transition time

$$\Delta t = \frac{l_e}{\bar{v}} = 4.67\,\text{s}$$

for the merging controller (cf. Fig. 7.18). Thus, the individual merging controller is initiated every time a robot on the merging lane enters the transition region at 3.42 m and the transition is finished at 5.52 m at the latest where the merging region begins. At this point, it is safe for the merging robots to change to the main lane. The negative inter-robot distances of the robots on different lanes can be explained with the fact that some of the robots on the merging lane overtake a robot on the main lane which is visible in the time interval [9 s, 13 s].

7.3.7 Summary of vehicle platooning

This section presented an experimental evaluation of the methods that have been developed in Chapter 4 to design adaptive cruise controllers. It has been shown and experimentally verified that external positivity of the controlled vehicle $\bar{\Sigma}_i$ is sufficient for collision

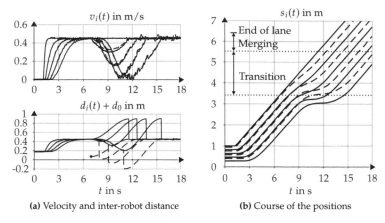

(a) Velocity and inter-robot distance **(b)** Course of the positions

Fig. 7.20: Merging of two platoons

avoidance in platoons of arbitrary length. Although $\mathcal{L}_2$ string stability is often preferred in the relevant literature, it does not guarantee collision avoidance as the experiments reveal.

As an extension to platooning, this section also showed an experimental evaluation of a concept for cooperative merging. The results show that the proposed methods perform properly in an experimental environment, i.e. the controllers allow for organising the robots in platoons and for merging of multiple platoons before a lane reduction. Furthermore, the control algorithms are real-time capable and achieve collision avoidance in all considered situations under the circumstance that there is no coordinating unit with overall knowledge. The proposed concept can also be generalised to control traffic situations with more complicated traffic junctions. As an outlook, the master thesis [14] considered an autonomous intersection using networked controllers without a coordinator.

7.4 Distance-based formation control

7.4.1 Swarms of mobile robots

The experiments in the previous sections considered mobile agents that are bound to prescribed paths as in daily traffic scenarios. The combination of lateral and longitudinal controllers has been used to evaluate methods for the control of autonomous vehicles. This section will abstract from that control paradigm in order to evaluate the methods developed in Chapters 5 and 6 concerning the control of swarms of mobile agents. Here, the robots are free to move on a driving surface without a path controller. The communication structure will be maintained to be a Delaunay triangulation at any time using Algorithm 5.4 and networked controllers synthesized with Algorithm 6.1 are applied to the robots to achieve a distance-based formation without any collisions.

7.4.2 Nominal test scenario

In the first experiment, the proposed parametrisation of the Morse potential function is evaluated in the critical case that is described in Section 6.3.4 where each edge has the desired length $d^\star$ except for one single edge that has the minimum allowed length d_0. The initial set-up and the formation to be taken are shown in Fig. 7.21. Given the parameters $d_0 = 0.18\,\text{m}$, $d^\star = 0.4\,\text{m}$, $u_{\max} = 1\,\text{m}\,\text{s}^{-1}$ and $N = 4$, the controller parameters are determined with (6.52) and (6.53) which result in

$$k_1 = 0.0105, \qquad\qquad k_2 = 5.6283.$$

Note that the desired distance $d^\star$ is chosen arbitrarily and *not* so as to satisfy (6.50) because of the following reasons. First, the test scenario is chosen so that the communication structure is fixed and, hence, there are no new edges that are established due to the Lawson flip (cf. Section 6.4.3). Second, the initial positions in Fig. 7.21 do not respect the condition (6.33) of feasible initial positions anyway. The purpose of the following measurement is to verify that a critically short edge does not get shorter.

The measurement of the overall potential $P(\boldsymbol{p}(t))$, the control inputs $u_i(t)$ and the distances $d_{ij}(t)$ is given in Fig. 7.22. The overall potential in the first graph starts close to the maximum value given by (6.51) due to the design goal (6.35) and decreases monotonically according to Lemma 6.2. The gap between the maximum value $3(N-2)k_1$ and the measured value is due to the fact that the configuration in Fig. 7.21 has $|\mathcal{E}| = 5$ out of six possible edges of a triangulation with $N = 4$ vertices. Note that the potential $P(\boldsymbol{p}(t))$ is continuous since Algorithm 5.4, which maintains the Delaunay triangulation, does not invoke any Lawson flip in this example.

The second graph shows the course of the control inputs of all robots which clearly respect the input limitation $u_{\max}$. The two robots that are very close to each other have a

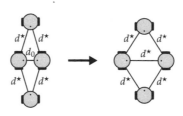

Fig. 7.21: Nominal test scenario with a critically short edge

control input that is shaped similarly to the course of the overall potential. In the beginning, it is close to the maximum value u_{max} and, then, decreases monotonically. The other two robot's control inputs are close to zero since they barely move at all (top and bottom in Fig. 7.21).

The robot distances $d_{ij}(t), (i, j) \in \mathcal{E}$ in the last graph approach the desired distance $d^\star$ asymptotically. In particular, the short edge with an initial length of d_0 expands without undershooting the minimum distance d_0. Thus, it can be concluded that no collision occurred despite the critically short edge. The controlled robots behave exactly as intended by the approach of the parametrisation of the Morse potential function.

7.4.3 Robot formation

The second experiment considers $N = 7$ robots that should approach the desired formation (6.21) starting from individual initial positions p_{i0} that satisfy (6.33) (Fig. 7.23, $t = 0\,$s) using Algorithm 6.1 as follows. With the parameters $d_0 = 0.18\,$m and $u_{\text{max}} = 1\,$m s^{-1} of the last example, the desired distance is set to $d^\star = 0.31\,$m using the largest value that satisfies (6.50). The coefficients of the Morse potential are determined with

$$k_1 = -\frac{(d_0 - d^\star)\, u_{\text{max}}}{2 \ln{(1 + \alpha\xi)}\,(1 + \alpha\xi)\,\alpha\xi} \tag{7.65}$$

$$k_2 = -\frac{\ln{(1 + \alpha\xi)}}{d_0 - d^\star}. \tag{7.66}$$

This choice of the parameters k_1 and k_2 respects the control input limitation u_{max}, i.e.

$$\frac{\mathrm{d}P_{ij}(d_0)}{\mathrm{d}d_{ij}(t)} = -u_{\text{max}}$$

still holds true. The equations (7.65) and (7.66) are obtained by introduction of the tuning factor α to (6.52) and (6.53) which is useful in practice due to the following reason. The robots used in the experiments do not execute very small control inputs. Due to the conservativity of the proposed solution, the resulting control actions are very small in

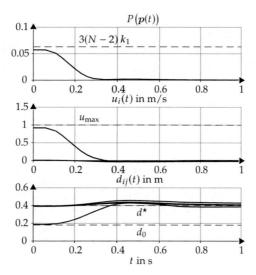

Fig. 7.22: Measurements of the test scenario in Fig. 7.21

most situations apart from the nominal test scenario in the last section. To overcome this limitation of the robots, a tuning factor $\alpha \in [0,1]$ reduces the number of edges virtually to obtain a controller parametrisation that is less conservative. The choice of $\alpha = 0.25$ leads to the controller parameters

$$k_1 = 0.0514, \qquad\qquad k_2 = 5.1071.$$

The measured course of the positions $p_i(t)$, $(i = 1, 2, \ldots, 7)$ to the final positions of the robots is shown in Fig. 7.23 with four intermediate points in time. The distance travelled to the current point in time is illustrated by the solid line in each frame. During the transition into the formation, the communication graph is adapted several times using the distributed algorithms presented in Chapter 5 so that it represents a Delaunay triangulation at any time. As an example, consider the edge $(3, 6)$ that has been established with Algorithm 5.3 in the third frame ($t = 7\,\text{s}$) due to the fact that agent 7 left the boundary of the convex hull where it was in the second frame ($t = 6\,\text{s}$). The very same edge $(3, 6)$ is removed in the last frame ($t = 25\,\text{s}$) by Algorithm 5.2 since agent 7 is part of the convex hull again. As the last example, consider the edge $(1, 3)$ which has been replaced by the edge $(4, 7)$ due to a Lawson flip that is invoked by Algorithm 5.1 during the transition from the second to last frame ($t = 12\,\text{s}$) to the last frame ($t = 25\,\text{s}$). The measurements reveal that the control aim

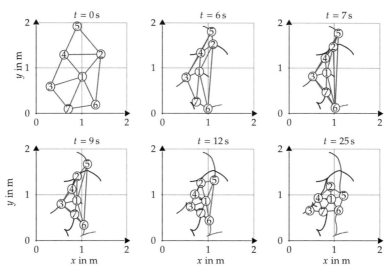

Fig. 7.23: Measured course of the positions $p_i(t)$

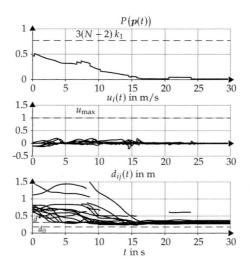

Fig. 7.24: Measurements of the formation in Fig. 7.23

is achieved, i. e. the robots form a distance-based formation where each edge has the same length approximately.

Similarly to the experiment in the last section, the measurements of the overall potential $P(\boldsymbol{p}(t))$, the control inputs $u_i(t)$ and the distances $d_{ij}(t)$ are given in Fig. 7.24. In contrast to the measurement in Fig. 7.22, the overall potential jumps at isolated points in time due to the switching character of the Delaunay triangulation. Nonetheless, the bound (6.51) is respected as shown by Theorem 6.3. Furthermore, it can be seen in the middle graph that the control inputs respect the limitation $u_{\max}$ and the distances exceed the minimum distance d_0 at any time so that collision avoidance can be concluded. Note that some curves are interrupted due to the switching characteristic of the Delaunay network. Whenever an edge gets removed or is established, a curve vanishes or appears, respectively.

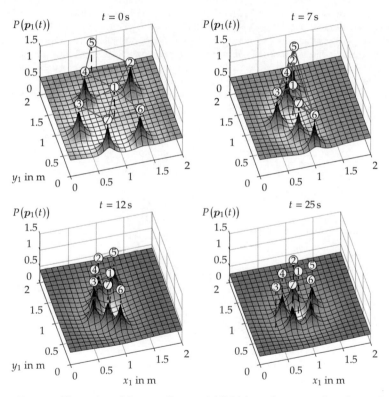

Fig. 7.25: Illustration of the overall potential field from the perspective of agent 1

As a supplement to the previous illustrations, Fig. 7.25 depicts the potential field from the perspective of agent 1 for four points in time. The potential field $P(p_1(t))$ is calculated as follows. For a given point in time, the sum (6.9) is evaluated for a variable candidate $p_1(t)$ and the current measurements of the positions $p_j, (j = 2, 3, \ldots, 7)$. The result is a hilly surface with elevations under the agents that are current neighbours of agent 1. The elevations result from the shape of the Morse potential function (cf. Fig. 6.5) which increases exponentially for small inter-robot distances. In the first frame ($t = 0\,\mathrm{s}$), the robots are located at their respective initial positions. Note that agent 5 has no elevation at its position in the potential field since it is not connected to agent 1, i.e. the edge $(1, 5)$ does not exist. From the perspective of agent 1, the movement of the robots into the distance-based formation displayed along the frames can be interpreted as a marble rolling into the lowest point between the elevations which is achieved by the gradient-based controller proposed in Section 6.2.3. During the transition into the stationary formation, all robot's movements result from the same principle and, hence, the shape of the potential field changes continuously. Furthermore, some elevations appear or disappear due to the dynamical adjustment of the communication network. The algorithms developed in Chapter 5 that maintain the Delaunay triangulation cause edges to be established or removed which can be observed in the transition from frame three ($t = 12\,\mathrm{s}$) to the final frame ($t = 25\,\mathrm{s}$) where the elevation at the position of agent 3 vanishes due to the fact that the edge $(1, 3)$ is removed by application of the Lawson flip.

7.4.4 Summary of distance-based formations

This section experimentally tested the results of the Chapters 5 and 6 with a set of mobile robots. The measurements have shown that the appropriate design of Morse potential functions can guarantee collision avoidance under consideration of control input limitations. The results have been combined with a switching proximity network in form of a Delaunay triangulation that is applied to the networked control system.

The results demonstrate that the Delaunay triangulation is an appropriate choice for the communication structure of a networked control system for the presented distance-based formation. The last experiment to be discussed in the next section will present another application for the Delaunay triangulation by extending the previous results to the case that the robots have to approach an individual destination simultaneously.

7.5 Transition problem

7.5.1 Introduction to the transition problem

This section addresses the transition problem where each robot should move from an initial position to an individual destination. The robots do not know where the other robots are heading and there is no initial trajectory planning. Thus, the robots have to navigate with local information and cooperate with other robots to avoid collisions if they come too close to each other. To this end, the robots use the communication network to exchange information about their current position with neighbouring robots. The control problem is formalised as follows.

Problem 7.2 (Transition control aim). Starting from the individual initial positions $p_i(0) = p_{i0}$, ($i \in \mathcal{V}$), all robots should approach their respective destination point $p_i^\star$ asymptotically:

$$\lim_{t \to \infty} \left\| p_i(t) - p_i^\star \right\| \stackrel{!}{=} 0, \quad i = 1, 2, \dots, N.$$

In order to avoid collisions throughout the transition to the destination points, the robots have to *cooperate* with each other using the communication network. As in the experiments presented in the last section, the communication structure is chosen to be the Delaunay triangulation.

If the inter-robot distance

$$d_{ij}(t) = \left\| p_i(t) - p_j(t) \right\|, \quad i \neq j \tag{7.67}$$

falls below the threshold δ, evasive manoeuvres have to be initiated. As a consequence, the set of feasible destination points

$$\mathcal{P}^\star := \{ p_1^\star, p_2^\star, \dots, p_N^\star \}$$

has to satisfy the condition

$$\left\| p_i^\star - p_j^\star \right\| \geq \delta, \quad \forall p_i^\star, p_j^\star \in \mathcal{P}^\star, \quad i \neq j,$$

which ensures that each destination point is separated at least by δ from the others.

7.5.2 Extended potential field

The transition problem will be solved using the gradient-based controller presented in Chapter 6 that minimises an artificial potential field. To this end, the overall potential defined by (6.9) is extended to consider the deviation of the individual destination of each robot. The result is the function

$$P(p(t)) := \sum_{i=1}^{N} \left(P_i^\star \left(\left\| p_i(t) - p_i^\star \right\| \right) + \frac{1}{2} \sum_{j \in \mathcal{N}_i} P_{ij}(d_{ij}(t)) \right), \tag{7.68}$$

in which each robot contributes with two types of potentials. On the one hand, it has the attracting potential $P_i^\star$ that penalises the deviation $\|p_i(t) - p_i^\star\|$ from its individual destination point. On the other hand, it has the repelling relative potentials P_{ij} that contribute to the overall potential if other robots $j \in \mathcal{N}_i$ are too close to robot i (cf. [51]). The potentials should be nonnegative, continuously differentiable and designed to meet the conditions

$$P_i^\star \left(\|p_i(t) - p_i^\star\| \right) = 0 \iff p_i(t) = p_i^\star \tag{7.69}$$

$$P_{ij}\big(d_{ij}(t)\big) = 0 \iff \|p_i(t) - p_j(t)\| \geq \delta, \tag{7.70}$$

respectively, i. e. $P\big(p(t)\big) = 0$ if and only if Problem 7.2 is solved. For a given potential field, the local controller that performs the gradient descent is given by (cf. (6.20))

$$F_i : \begin{pmatrix} u_i(t) \\ \omega_i(t) \end{pmatrix} = - \begin{pmatrix} e_i^\mathsf{T}(\phi_i(t)) \\ \mu\, n_i^\mathsf{T}(\phi_i(t)) \end{pmatrix} \nabla_{p_i} P\big(p(t)\big) \tag{7.71}$$

with the gradient

$$\nabla_{p_i} P\big(p(t)\big) = \nabla_{p_i} P_i^\star \left(\|p_i(t) - p_i^\star\| \right) + \sum_{j \in \mathcal{N}_i} \frac{\mathrm{d}P_{ij}\big(d_{ij}(t)\big)}{\mathrm{d}d_{ij}(t)} \frac{p_i(t) - p_j(t)}{d_{ij}(t)}$$

$$= \frac{\mathrm{d}P_i^\star \left(\|p_i(t) - p_i^\star\| \right)}{\mathrm{d}\|p_i(t) - p_i^\star\|} \frac{p_i(t) - p_i^\star}{\|p_i(t) - p_i^\star\|} + \sum_{j \in \mathcal{N}_i} \frac{\mathrm{d}P_{ij}\big(d_{ij}(t)\big)}{\mathrm{d}d_{ij}(t)} \frac{p_i(t) - p_j(t)}{d_{ij}(t)}. \tag{7.72}$$

Choice of the potential functions

The attracting potential is chosen as

$$P_i^\star \left(\|p_i(t) - p_i^\star\| \right) = v_\mathrm{T} \left(\|p_i(t) - p_i^\star\| - \alpha \ln \left(1 + \frac{\|p_i(t) - p_i^\star\|}{\alpha} \right) \right)$$

with the parameters v_T and α denoting the travelling velocity and a shaping factor, respectively. The derivative of the attracting potential with respect to the deviation from the destination is then given by

$$\frac{\mathrm{d}P_i^\star \left(\|p_i(t) - p_i^\star\| \right)}{\mathrm{d}\|p_i(t) - p_i^\star\|} = v_\mathrm{T} \frac{\|p_i(t) - p_i^\star\|}{\alpha + \|p_i(t) - p_i^\star\|}.$$

For large deviations $\|p_i(t) - p_i^\star\|$, the control input $u_i(t)$ is approximately equal to v_T under the control law (7.71) which can be used to take the control input bounds into account. The factor α allows for adjusting the behaviour of a robot close to its destination, e. g.

choosing a small value $\alpha > 0$ results in a reduction of the velocity as a robot approaches its destination. The repelling relative potential is chosen as

$$P_{ij}\big(d_{ij}(t)\big) = \begin{cases} k_R \dfrac{1}{2}\big(d_{ij}(t) - \delta\big)^2, & d_{ij}(t) \le \delta \\[2mm] 0, & \text{otherwise} \end{cases}$$

with the repelling gain $k_R > 0$ and the interaction threshold δ leading to the derivative

$$\frac{\mathrm{d}P_{ij}\big(d_{ij}(t)\big)}{\mathrm{d}d_{ij}(t)} = \begin{cases} k_R\big(d_{ij}(t) - \delta\big), & d_{ij}(t) \le \delta \\[2mm] 0, & \text{otherwise,} \end{cases} \tag{7.73}$$

which causes the robots to repel each other as soon as their distance falls below δ. Note that the potential functions are designed in accordance with the conditions (7.69) and (7.70).

With the presented choice of the attracting and repelling potential functions, the gradient (7.72) is explicitly given by

$$\nabla_{p_i} P\big(p(t)\big) = \begin{cases} \dfrac{v_T\big(p_i(t) - p_i^\star\big)}{\alpha + \big\|p_i(t) - p_i^\star\big\|} + \displaystyle\sum_{j \in \mathcal{N}_i} k_R \left(1 - \dfrac{\delta}{d_{ij}(t)}\right)\big(p_i(t) - p_j(t)\big), & d_{ij}(t) \le \delta \\[4mm] \dfrac{v_T\big(p_i(t) - p_i^\star\big)}{\alpha + \big\|p_i(t) - p_i^\star\big\|}, & \text{otherwise.} \end{cases} \tag{7.74}$$

7.5.3 Circle experiment

For the experimental evaluation, the robots will be placed evenly separated on a circle initially and the destination of each robot is on the opposite side of the circle as depicted exemplarily for robot $i = 1$ in Fig. 7.26 (a) (cf. [143] for a similar set-up). The initial condition is constructed as

$$x_{i0} = 2 + R \cos\left(\frac{2\pi(i - 1)}{N}\right) \tag{7.75}$$

$$y_{i0} = 2 + R \sin\left(\frac{2\pi(i - 1)}{N}\right) \tag{7.76}$$

$$\phi_{i0} = \frac{2\pi(i - 1)}{N} - \pi \tag{7.77}$$

so that the robots in the circular formation are oriented towards the centre of the circle. The parameters that fully characterise the feedback (7.71) with the gradient (7.74) are given in Table 7.3.

The first experiment is conducted with $N = 5$ robots and a circle formation with the radius $R = 1\,\mathrm{m}$. The graph in Fig. 7.26 (a) shows the evolution of the positions $p_i(t)$ of all robots starting from the initial position (7.75)–(7.77) depicted by circles o moving to

Table 7.3: Parameters

Parameter	μ	α	δ	k_R	v_T
Value	20	0.2 m	0.5 m	1.6	0.5 m/s

their individual destination $p_i^\star$ depicted by crosses ×. At first, each robot moves straight towards its destination. As soon as the robots meet in the centre and undercut the distance threshold δ, the repelling potentials (7.73) contribute to the gradient and cause evasive manoeuvres. Interestingly, all robots agree on turning simultaneously clockwise on a circular orbit to dodge each other and to reach their individual destination resulting in a well organised transition.

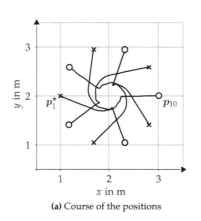

(a) Course of the positions

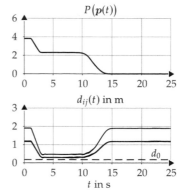

(b) Potential field and distances

Fig. 7.26: Circle experiment with $N = 5$ robots

Figure 7.26 (b) shows the course of the overall potential (7.68) and the inter-robot distances (7.67). The potential is monotonically decreasing since Lemma 6.2 still holds true with the extended potential function (7.68). There is a plateau in the time interval $t \in [3\,\text{s}, 10\,\text{s}]$ which can be explained as follows. When the robots meet close to the centre of the circle formation, the repelling potentials compensate for the attracting potentials. Thus, the resulting control input is rather small. Furthermore, as long as the robots turn in the circular orbit to evade each other, the distances remain constant approximately as can be observed in the bottom graph of Fig. 7.26 (b). Thus, the value of the potential hardly decreases during this evasive manoeuvre. Afterwards, the robots have free way to their individual destination and, hence, can complete the transition. The course of the distances furthermore reveals that the minimum distance $d_0 = 0.18\,\text{m}$ is exceeded at any point in time by all distances, i. e. collision avoidance is achieved in this experiment.

The measurement in Fig. 7.27 shows the same experiment with $N = 9$ robots on a circular formation with a radius of $R = 1.3\,\text{m}$. For the applied parameter values given in Table 7.3, there is no order in the transition as observed in the previous example. However, each robot arrives at its destination solving the control problem as shown by Fig. 7.27 (b) which illustrates the monotonically decreasing course of the overall potential that converges to zero asymptotically. Again, there are no collisions as shown by the course of the distances in the bottom of Fig. 7.27 (b) which are greater than the minimum distance d_0 at any time.

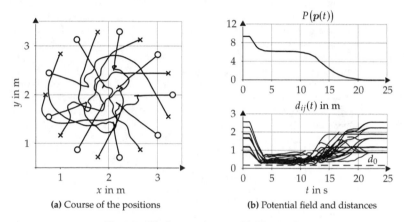

(a) Course of the positions   **(b)** Potential field and distances

Fig. 7.27: Circle experiment with $N = 9$ robots

Note that some of the distances are interrupted whenever an edge gets removed or is established due to the maintenance of the Delaunay triangulation network as has been already observed in the measurements in Fig. 7.24.

7.5.4 Summary of the transition problem

The combined tasks of collision avoidance and the transition to a destination can be achieved by an appropriate choice of the potential function which has been extended in this section to also consider the deviation of a robot from its destination point. The local feedback developed in Section 6.2.3 achieved a well organised cooperative behaviour of the robots during the transitions to their individual destination points for specific parameter values. Furthermore, the control algorithms are real-time capable and achieve collision avoidance in all considered situations under the circumstance that there is no coordinating unit with overall knowledge.

7.6   Summary and literature notes

This chapter presented several experiments concerning the cooperative control of networked vehicles on the basis of mobile robots. The experiments can be assigned to two classes. First, the combination of lateral and longitudinal controllers were discussed to study daily traffic scenarios that should be handled by autonomous vehicles. To this end, a path tracking controller presented in the beginning of this chapter has been developed to complement longitudinal controllers as the platooning controller proposed in Chapter 4 and the extension to merging of multiple platoons in the case of lane reductions with cooperative tracking controllers.

As the second class of control problem, swarms of mobile agents have been addressed that are allowed to move freely on a given driving surface. The theoretical fundament, which has been developed in Chapters 5 and 6, has been used in this chapter to test distance-based formations and to solve the transition problem.

The results in this chapter are based on the following contributions of the author. The path tracking controller has been developed in [10]. The comparative study concerning externally positive and $\mathcal{L}_2$ string stable vehicles in a platoon has been presented in [9]. The extension of the control concept to cooperative merging has been discussed in [2, 4, 13] theoretically and in [12] practically with the experiments that have been presented in this chapter. The evaluation of the distance-based formation has been published in the journal paper [7] with focus on the communication network and in [11] with focus on the collision avoidance. Finally, the transition problem together with the development of the gradient-based feedback has been published in [1].

Conclusion 8

The present thesis discussed different aspects concerning the cooperative control of networked vehicles. The collaboration of multiple vehicles, which are equipped with sensors and communication systems, has been studied as a networked control system. The general idea to derive properties of the controlled vehicles and properties of the communication network from requirements that describe the desired overall behaviour has served as a guideline throughout this thesis. The proposed results complement the existing literature with novel design procedures that satisfy specific design objectives and distributed algorithms to maintain a proximity communication structure based on the Delaunay triangulation. The most important contributions are summarised as follows.

Design of adaptive cruise controllers. The local controller of each vehicle has to achieve specific design goals so that the controlled vehicle behaves as desired in a situation that requires the cooperation of multiple vehicles. In context of vehicle platooning, the key property is external positivity which ensures collision-free transitions if the leader changes its velocity. In Chapter 3, it has been shown by Theorem 3.4 that plants with multiple integrators are not positively stabilisable with an output feedback which lead to the decision to use a state feedback structure for adaptive cruise control. The problem of choosing appropriate controller parameters has been solved using the pole assignment technique of J. Ackermann in Chapter 4. To this end, a method to find a set of closed-loop eigenvalues that correspond to an externally positive system is proposed. The eigenvalues should be placed by Ackermann's formula as summarised by Algorithm 4.2 which leads to an adaptive cruise controller with which the controlled vehicles can be combined in any order or number and satisfy the requirements on the overall platoon.

Maintenance of a Delaunay triangulation network. In a first step towards an abstraction from real traffic scenarios, the Delaunay triangulation has been proposed as the communication structure of a set of networked vehicles. The Delaunay triangulation has been shown to be a good choice for the communication structure for two reasons. First, it is characterised by local conditions which have been used to develop the distributed Algorithms 5.1–5.3 in Chapter 5 to be executed by the vehicles in order to maintain the network structure while the vehicles are moving. Thus, there is no need for a global coordinating unit. Second, the Delaunay triangulation yields a dense network with the guarantee that the two geometrically closest vehicles are always connected with each other.

163

This important property has been used to show that a Delaunay triangulation network in combination with an appropriate controller achieves collision avoidance.

Cooperative control of a swarm of networked vehicles. Distance-based formations have been studied in Chapter 6 as an application example for a swarm of networked vehicles. Due to the free movement of the vehicles on a driving surface that is not bound to specific paths, the nonholonomic kinematics of the vehicles had to be considered. For that, a gradient-based controller that realises a descent along an artificial potential field has been applied. It has been proposed to compose the overall field with Morse potential functions which allow for considering control input limitations due to the special shape of the potential. The parametrisation of the Morse potential function according to Theorem 6.1 has been proven to guarantee collision avoidance in combination with a Delaunay triangulation communication structure by Theorem 6.3.

Experimental evaluation

All theoretical results have been evaluated and verified in Chapter 7 with the laboratory set-up SAMS using a set of mobile robots with a differential drive. The proposed algorithms have been shown to be real-time capable and to work robustly in an experimental environment despite some parasitic effects as control input dead zones and limitations.

The evaluation of the longitudinal adaptive cruise controller required the development of a lateral path tracking controller that keeps the robots on a prescribed circular path and a longitudinal velocity controller for the leader. The combination of the lateral and longitudinal controllers has been used to compare externally positive vehicles with $\mathcal{L}_2$ string stable vehicles in a platoon. The design method proposed in Chapter 4 rendered the controlled vehicles externally positive, i. e. the assumptions (linear vehicle model, no transmission zeros) that have been imposed on the design procedure are realistic. As expected, the platoon with $\mathcal{L}_2$ string stable vehicles ended up in a collision in case of an emergency braking manoeuvre while the platoon with externally positive vehicles performed smoothly. As an outlook for further investigations concerning platoons of vehicles, an extension has been studied that allows for merging of multiple platoons in case of a lane reduction.

Further experiments have been conducted to test the methods concerning the vehicle swarm with mobile robots that are not bound to move on prescribed path in contrast to the previous experiments. The Delaunay triangulation has been applied as a communication structure in two experiments. First, the distance-based formation has been studied which showed that the robots approach the desired formation slowly but steadily using the Morse potential functions. In a second experiment, the overall potential has been extended to also consider the deviation of a robot to its destination point in order to study the circle transition problem. The experiments revealed a well ordered cooperative behaviour of the agents that first meet in the centre and, then, rotate clockwise on a circular orbit to dodge each other. Throughout the experiments, no collision has been observed.

Bibliography

Contributions of the Author

[1] Schwab, A., Littek, F., and Lunze, J. (2021). Experimental evaluation of a novel approach to cooperative control of multiple robots with artificial potential fields. *European Control Conference*, pp. 2388–2393.

[2] Schwab, A. and Lunze, J. (2019). Cooperative vehicle merging with guaranteed collision avoidance. *IFAC Workshop on Control of Transportation Systems*, pp. 7–12.

[3] Schwab, A. und Lunze, J. (2019). Kooperative Steuerung vernetzter Fahrzeuge. *VDI/VDE-Fachtagung AUTOREG*. Bd. VDI-Berichte Nr. 2349, S. 353–366.

[4] Schwab, A. and Lunze, J. (2019). Vehicle platooning and cooperative merging. *IFAC Symposium on Advances in Automotive Control*, pp. 353–358.

[5] Schwab, A. and Lunze, J. (2020). Design of adaptive cruise controllers for externally positive vehicles. *IFAC World Congress*, pp. 15275–15280.

[6] Schwab, A. and Lunze, J. (2020). How to design externally positive feedback loops - an open problem of control theory. *at-Automatisierungstechnik*, 68(5), pp. 301–311.

[7] Schwab, A. and Lunze, J. (2021). A distributed algorithm to maintain a proximity communication network among mobile agents using the Delaunay triangulation. *European Journal of Control*, 60, pp. 125–134.

[8] Schwab, A. and Lunze, J. (2021). Design of platooning controllers that achieve collision avoidance by external positivity. *IEEE Transactions on Intelligent Transportation Systems*, pp. 1–10.

[9] Schwab, A. and Lunze, J. (2021). On the collision avoidance of adaptive cruise controllers: Comparison of string stability and external positivity. *IEEE Conference on Control Technology and Applications*, pp. 20–25.

[10] Schwab, A. and Lunze, J. (2022). Analysis and experimental evaluation of a lateral controller for path tracking of mobile systems. *European Control Conference*, pp. 1997–2003.

[11] Schwab, A. and Lunze, J. (2022). Formation control over Delaunay triangulation networks with guaranteed collision avoidance. *IEEE Transactions on Control of Networked Systems*, (accepted).

[12] Schwab, A., Reichelt, L.-M., Welz, P., and Lunze, J. (2020). Experimental evaluation of an adaptive cruise control and cooperative merging concept. *IEEE Conference on Control Technology and Applications*, pp. 318–325.

[13] Schwab, A., Schenk, K., and Lunze, J. (2019). Networked vehicle merging by cooperative tracking control. *IFAC Workshop on Distributed Estimation and Control in Networked Systems*, pp. 19–24.

Supervised Theses

[14] Hinsen, P. (2021). *Ein Konzept zur verteilten Steuerung von Fahrzeugen an einer ampellosen Kreuzung*. Ruhr-Universität Bochum, Masterarbeit.

[15] Jansen, E. (2018). *Eine Methode zur Anpassung der Kommunikationsstruktur anhand von lokalen Regeln*. Ruhr-Universität Bochum, Masterarbeit.

[16] Littek, F. (2020). *Experimentelle Erprobung kooperativer Steuerungen für einen Roboterschwarm*. Ruhr-Universität Bochum, Masterarbeit.

[17] Reichelt, L.-M. (2020). *Erprobung einer Methode zur kooperativen Einfädelung von Fahrzeugen vor einer Fahrbahnzusammenführung*. Ruhr-Universität Bochum, Masterarbeit.

Further Literature

[18] Ackermann, J. (1972). Der Entwurf linearer Regelungssysteme im Zustandsraum. *at-Automatisierungstechnik*, 7, S. 297–300.

[19] Albers, G., Guibas, L. J., Mitchell, J. S., and Roos, T. (1998). Voronoi diagrams of moving points. *International Journal of Computational Geometry & Applications*, 8(03), pp. 365–379.

[20] Anderson, S. J., Karumanchi, S. B., and Iagnemma, K. (2012). Constraint-based planning and control for safe, semi-autonomous operation of vehicles. *IEEE Intelligent Vehicles Symposium*, pp. 383–388.

[21] Ando, H., Oasa, Y., Suzuki, I., and Yamashita, M. (1999). Distributed memory-less point convergence algorithm for mobile robots with limited visibility. *IEEE Transactions on Robotics and Automation*, 15(5), pp. 818–828.

[22] Attia, R., Orjuela, R., and Basset, M. (2014). Combined longitudinal and lateral control for automated vehicle guidance. *Vehicle System Dynamics*, 52(2), pp. 261–279.

[23] Benner, P., Gugercin, S., and Willcox, K. (2015). A survey of projection-based model reduction methods for parametric dynamical systems. *SIAM Review*, 57(4), pp. 483–531.

[24] Berman, A. and Plemmons, R. J. (1994). *Nonnegative Matrices in the Mathematical Sciences*. SIAM.

[25]   Bianchini, F., Samaniego, C. C., Franco, E., and Giordano, G. (2017). Aggregates of positive impulse response systems: A decomposition approach for complex networks. *IEEE Conference on Decision and Control*, pp. 1987–1992.

[26]   Bogachev, V. I. (2007). *Measure Theory*. Springer Science & Business Media.

[27]   Bolzern, P. and Colaneri, P. (2015). *Positive Markov Jump Linear Systems*. Now Publishers Inc.

[28]   Brandão, A. S. and Sarcinelli-Filho, M. (2016). On the guidance of multiple UAV using a centralized formation control scheme and Delaunay triangulation. *Journal of Intelligent & Robotic Systems*, 84(1-4), pp. 397–413.

[29]   Branicky, M. S. (1994). Stability of switched and hybrid systems. *IEEE Conference on Decision and Control*, pp. 3498–3503.

[30]   Branicky, M. S. (1998). Multiple Lyapunov functions and other analysis tools for switched and hybrid systems. *IEEE Transactions on Automatic Control*, 43(4), pp. 475–482.

[31]   Cortés, J., Martínez, S., and Bullo, F. (2006). Robust rendezvous for mobile autonomous agents via proximity graphs in arbitrary dimensions. *IEEE Transactions on Automatic Control*, 51(8), pp. 1289–1298.

[32]   Cortés, J., Martínez, S., Karatas, T., and Bullo, F. (2004). Coverage control for mobile sensing networks. *IEEE Transactions on Robotics and Automation*, 20(2), pp. 243–255.

[33]   De Berg, M., Van Kreveld, M., Overmars, M., and Schwarzkopf, O. (2008). *Computational Geometry*. Springer.

[34]   DeCarlo, R. A., Branicky, M. S., Pettersson, S., and Lennartson, B. (2000). Perspectives and results on the stability and stabilizability of hybrid systems. *Proceedings of the IEEE*, 88(7), pp. 1069–1082.

[35]   Delaunay, B. (1934). Sur la sphere vide. *Izv. Akad. Nauk SSSR, Otdelenie Matematicheskii i Estestvennyka Nauk*, 7(793-800), pp. 1–2.

[36]   Dimarogonas, D. V. and Johansson, K. H. (2008). On the stability of distance-based formation control. *IEEE Conference on Decision and Control*, pp. 1200–1205.

[37]   Eyre, J., Yanakiev, D., and Kanellakopoulos, I. (1998). A simplified framework for string stability analysis of automated vehicles. *Vehicle System Dynamics*, 30(5), pp. 375–405.

[38]   Falconi, R., Sabattini, L., Secchi, C., Fantuzzi, C., and Melchiorri, C. (2011). A graph-based collision-free distributed formation control strategy. *IFAC World Congress*, pp. 6011–6016.

[39]   Farina, L. and Rinaldi, S. (2011). *Positive Linear Systems: Theory and Applications*. John Wiley & Sons.

[40]   Fax, J. A. and Murray, R. M. (2004). Information flow and cooperative control of vehicle formations. *IEEE Transactions on Automatic Control*, 49(9), pp. 1465–1476.

[41]   Fu, J.-J. and Lee, R. C. T. (1991). Voronoi diagrams of moving points in the plane. *International Journal of Computational Geometry & Applications*, 1(01), pp. 23–32.

[42]   Garone, E. and Ntogramatzidis, L. (2015). Linear matrix inequalities for globally monotonic tracking control. *Automatica*, 61, pp. 173–177.

[43] Glover, J. (1900). Transition curves for railways. *Minutes of the Proceedings of the Institution of Civil Engineers*. Vol. 140. 1900, pp. 161–179.

[44] Groß, D. and Stursberg, O. (2011). Optimized distributed control and network topology design for interconnected systems. *IEEE Conference on Decision and Control*, pp. 8112–8117.

[45] Guibas, L. and Stolfi, J. (1985). Primitives for the manipulation of general subdivisions and the computation of Voronoi diagrams. *ACM Transactions on Graphics*, 4(2), pp. 74–123.

[46] Gunter, G., Janssen, C., Barbour, W., Stern, R. E., and Work, D. B. (2020). Model-based string stability of adaptive cruise control systems using field data. *IEEE Transactions on Intelligent Vehicles*, 5(1), pp. 90–99.

[47] Gunter, G. et al. (2020). Are commercially implemented adaptive cruise control systems string stable? *IEEE Transactions on Intelligent Transportation Systems*,

[48] Gusrialdi, A. and Hirche, S. (2010). Performance-oriented communication topology design for large-scale interconnected systems. *IEEE Conference on Decision and Control*, pp. 5707–5713.

[49] Haddad, W. M., Chellaboina, V., and Hui, Q. (2010). *Nonnegative and Compartmental Dynamical Systems*. Princeton University Press.

[50] Han, D. and Panagou, D. (2019). Robust multitask formation control via parametric Lyapunov-like barrier functions. *IEEE Transactions on Automatic Control*, 64(11), pp. 4439–4453.

[51] Hernandez-Martinez, E. G. and Aranda-Bricaire, E. (2011). "Convergence and collision avoidance in formation control: A survey of the artificial potential functions approach". *Multi-Agent Systems: Modeling, Control, Programming, Simulations and Applications*. Ed. by F. Alkhateeb et al. InTech. Chap. 6, pp. 103–126.

[52] Hjelle, Ø. and Dæhlen, M. (2006). *Triangulations and Applications*. Springer Science & Business Media.

[53] Horowitz, I. M. (1963). *Synthesis of Feedback Systems*. Academic Press.

[54] Jadbabaie, A., Lin, J., and Morse, A. S. (2003). Coordination of groups of mobile autonomous agents using nearest neighbor rules. *IEEE Transactions on automatic control*, 48(6), pp. 988–1001.

[55] Jarvis, R. A. (1973). On the identification of the convex hull of a finite set of points in the plane. *Information Processing Letters*, 2(1), pp. 18–21.

[56] Jayasuriya, S. and Franchek, M. (1991). A class of transfer functions with non-negative impulse response. *ASME Journal of Dynamic Systems, Measurement and Control*, 113(2), pp. 313–315.

[57] Jayasuriya, S. and Dharne, A. G. (2002). Necessary and sufficient conditions for non-overshooting step responses for LTI systems. *IEEE American Control Conference*. Vol. 1, pp. 505–510.

[58] Jayasuriya, S. and Song, J.-W. (1992). On the synthesis of compensators for non-overshooting step response. *American Control Conference*, pp. 683–684.

[59] Jia, Y., Jibrin, R., and Görges, D. (2020). Energy-optimal adaptive cruise control for electric vehicles based on linear and nonlinear model predictive control. *IEEE Transactions on Vehicular Technology*, 69(12), pp. 14173–14187.

[60] Jia, Y., Jibrin, R., Itoh, Y., and Görges, D. (2019). Energy-optimal adaptive cruise control for electric vehicles in both time and space domain based on model predictive control. *IFAC Symposium on Advances in Automotive Control*, pp. 13–20.

[61] Kaczorek, T. (2012). *Positive 1D and 2D Systems*. Springer Science & Business Media.

[62] Kalman, R. E. (1960). Contributions to the theory of optimal control. *Boletin de la Sociedad Matematica Mexicana*, 5(2), pp. 102–119.

[63] Katriniok, A. and Abel, D. (2016). Adaptive EKF-based vehicle state estimation with online assessment of local observability. *IEEE Transactions on Control Systems Technology*, 24(4), pp. 1368–1381.

[64] Kayacan, E. (2017). Multiobjective H_∞ control for string stability of cooperative adaptive cruise control systems. *IEEE Transactions on Intelligent Vehicles*, 2(1), pp. 52–61.

[65] Keil, J. M. and Gutwin, C. A. (1992). Classes of graphs which approximate the complete Euclidean graph. *Discrete & Computational Geometry*, 7(1), pp. 13–28.

[66] Khalil, H. K. (2002). *Nonlinear Systems*. Prentice Hall.

[67] El-Khoury, M., Crisalle, O. D., and Longchamp, R. (1993). Influence of zero locations on the number of step-response extrema. *Automatica*, 29(6), pp. 1571–1574.

[68] Klinge, S. and Middleton, R. H. (2009). Time headway requirements for string stability of homogeneous linear unidirectionally connected systems. *IEEE Conference on Decision and Control*, pp. 1992–1997.

[69] Koditschek, D. E. and Rimon, E. (1990). Robot navigation functions on manifolds with boundary. *Advances in Applied Mathematics*, 11, p. 412.

[70] Lawson, C. L. (1977). "Software for C1 surface interpolation". *Mathematical Software*. Ed. by J. R. Rice. Elsevier, pp. 161–194.

[71] Lawson, C. L. (1972). Transforming triangulations. *Discrete Mathematics*, 3(4), pp. 365–372.

[72] Lee, D.-T. and Schachter, B. J. (1980). Two algorithms for constructing a Delaunay triangulation. *International Journal of Computer & Information Sciences*, 9(3), pp. 219–242.

[73] Levine, W. and Athans, M. (1966). On the optimal error regulation of a string of moving vehicles. *IEEE Transactions on Automatic Control*, 11(3), pp. 355–361.

[74] Li, G. and Görges, D. (2018). Ecological adaptive cruise control and energy management strategy for hybrid electric vehicles based on heuristic dynamic programming. *IEEE Transactions on Intelligent Transportation Systems*, 20(9), pp. 3526–3535.

[75] Li, X.-Y., Calinescu, G., Wan, P.-J., and Wang, Y. (2003). Localized Delaunay triangulation with application in ad hoc wireless networks. *IEEE Transactions on Parallel and Distributed Systems*, 14(10), pp. 1035–1047.

[76] Liberzon, D. (2003). *Switching in Systems and Control*. Springer Science & Business Media.

[77] Listmann, K. D., Masalawala, M. V., and Adamy, J. (2009). Consensus for formation control of nonholonomic mobile robots. *IEEE International Conference on Robotics and Automation*, pp. 3886–3891.

[78] Liu, Y. and Bauer, P. H. (2008). Sufficient conditions for non-negative impulse response of arbitrary-order systems. *IEEE Asia Pacific Conference on Circuits and Systems*, pp. 1410–1413.

[79] Luenberger, D. G. (1979). *Introduction to Dynamic Systems*. John Wiley & Sons.

[80] Lunze, J. (1992). *Feedback Control of Large-Scale Systems*. Prentice Hall.

[81] Lunze, J. (2011). An internal-model principle for the synchronisation of autonomous agents with individual dynamics. *IEEE Conference on Decision and Control*, pp. 2106–2111.

[82] Lunze, J. (2012). Synchronization of heterogeneous agents. *IEEE Transactions on Automatic Control*, 57(11), pp. 2885–2890.

[83] Lunze, J. (2013). A method for designing the communication structure of networked controllers. *International Journal of Control*, 86(9), pp. 1489–1502.

[84] Lunze, J. (2019). Adaptive cruise control with guaranteed collision avoidance. *IEEE Transactions on Intelligent Transportation Systems*, 20(5), pp. 1897–1907.

[85] Lunze, J. (2019). Design of the communication structure of cooperative adaptive cruise controllers for vehicle platoons. *IFAC Symposium on Advances in Automotive Control*, pp. 7–12.

[86] Lunze, J. (2019). *Networked Control of Multi-Agent Systems*. Bookmundo Direct.

[87] Lunze, J. (2020). *Automatisierungstechnik*. De Gruyter.

[88] Lunze, J. (2020). Design of the communication structure of cooperative adaptive cruise controllers. *IEEE Transactions on Intelligent Transportation Systems*, 21(10), pp. 4378–4387.

[89] Lunze, J. (2020). *Regelungstechnik 1*. Springer.

[90] Lunze, J. (2020). *Regelungstechnik 2*. Springer.

[91] Lunze, J. and Lamnabhi-Lagarrigue, F. (2009). *Handbook of Hybrid Systems Control: Theory, Tools, Applications*. Cambridge University Press.

[92] Lunze, J. and Lehmann, D. (2010). A state-feedback approach to event-based control. *Automatica*, 46(1), pp. 211–215.

[93] Mahmood, A. and Kim, Y. (2014). Collision-free vehicle formation control using graph Laplacian and edge-tension function. *IFAC World Congress*, pp. 1808–1812.

[94] Mastellone, S., Stipanović, D. M., Graunke, C. R., Intlekofer, K. A., and Spong, M. W. (2008). Formation control and collision avoidance for multi-agent non-holonomic systems: Theory and experiments. *The International Journal of Robotics Research*, 27(1), pp. 107–126.

[95] Mathieson, L. and Moscato, P. (2019). "An introduction to proximity graphs". *Business and Consumer Analytics: New Ideas*. Ed. by P. Moscato and N. J. de Vries. Springer. Chap. 4, pp. 213–233.

[96] Meguerdichian, S., Koushanfar, F., Potkonjak, M., and Srivastava, M. B. (2001). Coverage problems in wireless ad-hoc sensor networks. *IEEE Conference on Computer Communications*. Vol. 3, pp. 1380–1387.

[97] Mesbahi, M. and Egerstedt, M. (2010). *Graph Theoretic Methods in Multiagent Networks*. Princeton University Press.

[98] Milanés, V. and Shladover, S. E. (2014). Modeling cooperative and autonomous adaptive cruise control dynamic responses using experimental data. *Transportation Research Part C: Emerging Technologies*, 48, pp. 285–300.

[99] Milanés, V. et al. (2014). Cooperative adaptive cruise control in real traffic situations. *IEEE Transactions on Intelligent Transportation Systems*, 15(1), pp. 296–305.

[100] Moler, C. and Van Loan, C. (2003). Nineteen dubious ways to compute the exponential of a matrix, twenty-five years later. *SIAM review*, 45(1), pp. 3–49.

[101] Morse, P. M. (1929). Diatomic molecules according to the wave mechanics. II. Vibrational levels. *Physical Review*, 34(1), p. 57.

[102] Mosebach, A., Röchner, S., and Lunze, J. (2016). Merging control of cooperative vehicles. *IFAC Symposium on Advances in Automotive Control*, 49(11), pp. 168–174.

[103] Naus, G. J. L., Vugts, R. P. A., Ploeg, J., De Molengraft, M. J. G. van, and Steinbuch, M. (2010). String-stable CACC design and experimental validation: A frequency-domain approach. *IEEE Transactions on Vehicular Technology*, 59(9), pp. 4268–4279.

[104] Nunen, E. van, Reinders, J., Semsar-Kazerooni, E., and Van De Wouw, N. (2019). String stable model predictive cooperative adaptive cruise control for heterogeneous platoons. *IEEE Transactions on Intelligent Vehicles*, 4(2), pp. 186–196.

[105] Oh, K.-K., Park, M.-C., and Ahn, H.-S. (2015). A survey of multi-agent formation control. *Automatica*, 53, pp. 424–440.

[106] Olfati-Saber, R. (2006). Flocking for multi-agent dynamic systems: Algorithms and theory. *IEEE Transactions on Automatic Control*, 51(3), pp. 401–420.

[107] Olfati-Saber, R., Fax, J. A., and Murray, R. M. (2007). Consensus and cooperation in networked multi-agent systems. *Proceedings of the IEEE*, 95(1), pp. 215–233.

[108] Olfati-Saber, R. and Murray, R. M. (2002). Distributed cooperative control of multiple vehicle formations using structural potential functions. *IFAC World Congress*, pp. 242–248.

[109] Olfati-Saber, R. and Murray, R. M. (2004). Consensus problems in networks of agents with switching topology and time-delays. *IEEE Transactions on Automatic Control*, 49(9), pp. 1520–1533.

[110] Ollero, A. and Heredia, G. (1995). Stability analysis of mobile robot path tracking. *IEEE Conference on Intelligent Robots and Systems. Human Robot Interaction and Cooperative Robots*. Vol. 3, pp. 461–466.

[111] Oppenheim, A. V. and Willsky, A. S. (1983). *Signals and Systems*. Prentice Hall.

[112] Panagou, D. (2017). A distributed feedback motion planning protocol for multiple unicycle agents of different classes. *IEEE Transactions on Automatic Control*, 62(3), pp. 1178–1193.

[113] Panagou, D., Stipanović, D. M., and Voulgaris, P. G. (2016). Distributed coordination control for multi-robot networks using Lyapunov-like barrier functions. *IEEE Transactions on Automatic Control*, 61(3), pp. 617–632.

[114] Ploeg, J., Scheepers, B. T., Van Nunen, E., Van de Wouw, N., and Nijmeijer, H. (2011). Design and experimental evaluation of cooperative adaptive cruise control. *IEEE Conference on Intelligent Transportation Systems*, pp. 260–265.

[115] Ploeg, J., Van De Wouw, N., and Nijmeijer, H. (2013). $\mathcal{L}_p$ string stability of cascaded systems: Application to vehicle platooning. *IEEE Transactions on Contr. Systems Technology*, 22(2), pp. 786–793.

[116] Preparata, F. P. and Shamos, M. I. (2012). *Computational Geometry: An Introduction*. Springer Science & Business Media.

[117] Rachid, A. (1995). Some conditions on zeros to avoid step-response extrema. *IEEE Transactions on Automatic Control*, 40(8), pp. 1501–1503.

[118] Rantzer, A. and Valcher, M. E. (2018). A tutorial on positive systems and large scale control. *IEEE Conference on Decision and Control*, pp. 3686–3697.

[119] Regmi, A., Sandoval, R., Byrne, R., Tanner, H., and Abdallah, C. (2005). Experimental implementation of flocking algorithms in wheeled mobile robots. *American Control Conference*, pp. 4917–4922.

[120] Ren, W. and Beard, R. W. (2005). Consensus seeking in multiagent systems under dynamically changing interaction topologies. *IEEE Transactions on Automatic Control*, 50(5), pp. 655–661.

[121] Ren, W. and Beard, R. W. (2008). *Distributed Consensus in Multi-Vehicle Cooperative Control*. Springer.

[122] Ren, W., Beard, R. W., and Atkins, E. M. (2007). Information consensus in multivehicle cooperative control. *IEEE Control Systems Magazine*, 27(2), pp. 71–82.

[123] Rimon, E. and Koditschek, D. E. (1992). Exact robot navigation using artificial potential functions. *IEEE Transactions on Robotics and Automation*, 8(5), pp. 501–518.

[124] Samson, C. (1992). Path following and time-varying feedback stabilization of a wheeled mobile robot. *International Conference on Control, Automation, Robotics and Vision*.

[125] Scardovi, L. and Sepulchre, R. (2008). Synchronization in networks of identical linear systems. *IEEE Conference on Decision and Control*, pp. 546–551.

[126] Schmid, R. and Ntogramatzidis, L. (2010). A unified method for the design of nonovershooting linear multivariable state-feedback tracking controllers. *Automatica*, 46(2), pp. 312–321.

[127] Seiler, P., Pant, A., and Hedrick, K. (2004). Disturbance propagation in vehicle strings. *IEEE Transactions on Automatic Control*, 49(10), pp. 1835–1842.

[128] Semsar-Kazerooni, E., Verhaegh, J., Ploeg, J., and Alirezaei, M. (2016). Cooperative adaptive cruise control: An artificial potential field approach. *IEEE Intelligent Vehicles Symposium*, pp. 361–367.

[129] Shevitz, D. and Paden, B. (1994). Lyapunov stability theory of nonsmooth systems. *IEEE Transactions on Automatic Control*, 39(9), pp. 1910–1914.

[130] Shewchuk, R. (2005). Star splaying: An algorithm for repairing Delaunay triangulations and convex hulls. *Symposium on Computational Geometry*, pp. 237–246.

[131] Shin, D. H., Singh, S., and Lee, J. J. (1992). Explicit path tracking by autonomous vehicles. *Robotica*, 10(6), pp. 539–554.

[132] Swaroop, D. and Hedrick, J. K. (1996). String stability of interconnected systems. *IEEE Transactions on Automatic Control*, 41(3), pp. 349–357.

[133] Swaroop, D. and Niemann, D. (1996). Some new results on the oscillatory behavior of impulse and step responses for linear time invariant systems. *IEEE Conference on Decision and Control*. Vol. 3, pp. 2511–2512.

[134] Swaroop, D. (2003). On the synthesis of controllers for continuous time LTI systems that achieve a non-negative impulse response. *Automatica*, 39(1), pp. 159–165.

[135] Tanner, H. G., Jadbabaie, A., and Pappas, G. J. (2003). Stable flocking of mobile agents part II: Dynamic topology. *IEEE Conference on Decision and Control*. Vol. 2, pp. 2016–2021.

[136] Tanner, H. G., Jadbabaie, A., and Pappas, G. J. (2003). Stable flocking of mobile agents, part I: Fixed topology. *IEEE Conference on Decision and Control*. Vol. 2, pp. 2010–2015.

[137] Tanner, H. G., Jadbabaie, A., and Pappas, G. J. (2007). Flocking in fixed and switching networks. *IEEE Transactions on Automatic control*, 52(5), pp. 863–868.

[138] Valcher, M. E. and Misra, P. (2013). On the stabilizability and consensus of positive homogeneous multi-agent dynamical systems. *IEEE Transactions on Automatic Control*, 59(7), pp. 1936–1941.

[139] Van Arem, B., Van Driel, C. J., and Visser, R. (2006). The impact of cooperative adaptive cruise control on traffic-flow characteristics. *IEEE Transactions on Intelligent Transportation Systems*, 7(4), pp. 429–436.

[140] Verginis, C. K., Bechlioulis, C. P., Dimarogonas, D. V., and Kyriakopoulos, K. J. (2018). Robust distributed control protocols for large vehicular platoons with prescribed transient and steady-state performance. *IEEE Transactions on Control Systems Technology*, 26(1), pp. 299–304.

[141] Vicsek, T., Czirók, A., Ben-Jacob, E., Cohen, I., and Shochet, O. (1995). Novel type of phase transition in a system of self-driven particles. *Physical Review Letters*, 75(6), p. 1226.

[142] Vinberg, E. B. (2003). *A Course in Algebra*. American Mathematical Society.

[143] Wang, L., Ames, A. D., and Egerstedt, M. (2017). Safety barrier certificates for collisions-free multirobot systems. *IEEE Transactions on Robotics*, 33(3), pp. 661–674.

[144] Wang, Q., Chen, Z., Liu, P., and Hua, Q. (2017). Distributed multi-robot formation control in switching networks. *Neurocomputing*, 270, pp. 4–10.

[145] Welz, P., Fischer, M. und Lunze, J. (2020). Experimentelle Erprobung einer kollisionsfreien Abstandsregelung für mobile Roboter. *at-Automatisierungstechnik*, 68(1), S. 32–43.

[146]   Xiao, L. and Gao, F. (2011). Practical string stability of platoon of adaptive cruise control vehicles. *IEEE Transactions on Intelligent Transportation Systems*, 12(4), pp. 1184–1194.

[147]   Zhang, W., Branicky, M. S., and Phillips, S. M. (2001). Stability of networked control systems. *IEEE Control Systems Magazine*, 21(1), pp. 84–99.

[148]   Zhou, Y., Sun, F., Wang, W., Wang, J., and Zhang, C. (2010). Fast Updating of Delaunay triangulation of moving points by bi-cell filtering. *Computer Graphics Forum*, 29(7), pp. 2233–2242.

List of Symbols

The following tables list some important symbols that are used in this thesis. Symbols that characterise the closed-loop or the open-loop system have a bar or a subscript zero, respectively (e. g. $\bar{A}$ or A_0). Further notes on the notation are given in Section 2.1.

Important symbols 1	
A, b, c^T	Elements of a state-space model
β	Time-headway coefficient
C_i	Local controller
$d_i(t)$	Inter-vehicle distance (platooning)
$d_{ij}(t)$	Inter-vehicle distance (swarming)
$d^\star, d_0$	Desired distance, minimum distance
D_i	Decision and communication unit
$\delta(t)$	Dirac delta function
Δ	Delay measure
$e(t)$	Control error
$e_i(\phi_i(t))$	Orientation vector
$\mathcal{E}$	Set of edges
F_i	Feedback unit
$g(t)$	Impulse response
$G(s)$	Transfer function
$\mathcal{G}$	Graph
$h(t)$	Step response
I	Identity matrix
$k, k_\mathrm{P}, k_\mathrm{I}$	Feedback gain
$K(s)$	Controller transfer function
L	Laplacian matrix
$\lambda_i, \lambda_\mathrm{dom}$	Eigenvalue of A, dominant eigenvalue
m_{ABC}	Circumcenter
M_{ABCD}	Enclosure matrix
μ_i	Invariant zero

Important symbols 2	
n	System order
$\boldsymbol{n}_i(\phi_i(t))$	Normal vector
N	Number of agents
$\mathcal{N}_i$	Set of neighbours
$\boldsymbol{O}$	Zero matrix
$\boldsymbol{O}_{ABC}$	Orientation matrix
$\boldsymbol{p}_i(t)$	Position vector
P_i	Plant model
$P(\lambda)$	Characteristic polynomial
$\boldsymbol{P}(s)$	Rosenbrock system matrix
$P(\boldsymbol{p}(t))$	Artificial potential field
$P_{ij}(d_{ij}(t))$	Relative potential function
$\phi_i(t)$	Orientation angle
q	Number of zeros of a system
r	Relative degree
R	Radius
$s_i(t)$	Longitudinal position
s_i	Pole of $G(s)$
s_{0i}	Transmission zero of $G(s)$
$\boldsymbol{S}$	Controllability matrix
$\sigma(t)$	Heaviside step function
Σ, Σ_i	Isolated system, vehicle model
τ	Time constant, time lag
$u(t)$	System input, control input
$v_i(t), v_0(t)$	Translational velocity, reference of the leader
$\mathcal{V}$	Set of vertices
$w(t)$	Reference input, set-point
$\omega_i(t)$	Rotational velocity
$\boldsymbol{x}(t), \boldsymbol{x}_0$	System state, initial state
$x_i(t), y_i(t)$	Cartesian coordinates
$y(t)$	System output
$z_i(t)$	Regulator state